OKT

Forevermore

GREENTREES

THESE LETTERS ARE ALL PATTERNED HEIROGLYPHS

THEY ARE ALSO SHAPES & INVOKE SOUND (FREQUENCY)

THE FUTURE IS NOW

CONTENTS

GO BACK

REMEMBER

GREENTREES ACADEMY

$$
\begin{array}{r}
2600 \\
\underline{2020} \\
580
\end{array}
$$

$$
\begin{array}{rrrr}
2551 & 2551 & 2551 & 1980 \\
\underline{1980} & \underline{1836} & \underline{1971} & \underline{1836} \\
571 & 715 & 580 & 144
\end{array}
$$

$$
\begin{array}{rrrr}
1964 & 2000 & 1943 & 1953 \\
\underline{1906} & \underline{1964} & \underline{1906} & \underline{1809} \\
58 & 36 & 36 & 144
\end{array}
$$

Khufu - Tesla - Brett:
2566   1856-1943

2000	144	1856	2
	144	1712	2
	144	1568	2
	144	1424	2
	144	1280	2
	144	1136	2
	144	992	2
	144	848	2
	144	704	2
	144	560	2
	144	416	2
	144	272	2
	144	128	2
	144	**-16**	**7**
	144	**-160**	**7**
	144	-304	7
	144	-448	7
	144	-592	7
	144	-736	7
	144	-880	7

Menkaure - Einstein - Breinstein:
2532-2500 1897-1955

2016	144	1872	9
	144	1728	9
	144	1584	9
	144	**1440**	9
	144	1296	9
	144	1152	9
	144	1008	9
	144	864	9
	144	720	9
	144	576	9
	144	432	9
	144	**288**	9
	144	**144**	9
	144	0	0-9
	144	**-144**	9
	144	**-288**	9
	144	**-432**	9
	144	**-576**	9
	144	**-720**	9
	144	**-864**	9

Hetepheres - Sarah - Lacey/Genie:
2600-2551   1836-1922

1980	144	1836	9
	144	1692	9
	144	1548	9
	144	1404	9
	144	1260	9
	144	1116	9
	144	972	9
	144	828	9
	144	684	9
	144	540	9
	144	396	9
	144	252	9
	144	**108**	9
	144	-36	9
	144	**-180**	9
	144	-324	9
	144	-468	9
	144	-612	9
	144	-756	9
	144	-900	9

2000 **Brett b.**   1980 **Lacey b.**   1836 **Sarah b.**
<u>1856</u> **Tesla b.**   <u>1836</u> **Sarah b.**   <u>1692</u> **Salem Witch Trials**
 144   144   144

2585 **Khufu b.**   2585 **Khufu b.**   729
<u>2000</u> **Brett b.**   <u>1856</u> **Tesla b.**   <u>585</u>
 585   729   144

1953 **Rick Allan Moranis b.**
<u>1809</u> **Edgar Allan Poe b.**
 144

Meidum Pyramid: Base 144 metres (472 ft)

2600 - Hetepheres b.
2551 - Hetepheres d.
2500 - Khafre built Valley
2492 - ?
1809 - Poe b.
1836 - Sarah b.
1840 - Poe d.
1906 - Experiment-Earthquake
1912 - Titanic
1922 - Sarah d.
1926 - Queen Bee Interview
1943 - Tesla d.
1953 - Rick Moranis b.
1964 - Dog Letter
1971 - Elon b.
1980 - Lacey b.
2000 - Brett b.
2020 - Now
2022 - Dog Letter Comes Alive

KEY DATES

FUN FACTS

The new Khufu-Tesla likes spaghetti, nuggets, soft pretzels, veggie sausage, root beer, brownies, pizza, chocolate chip cookies, coffee, vanilla ice cream. He is connected to wood-trees. Green & blue. Sitting. Playing. Telling stories.   His father's birthday is the same as his previous father's death date: **4-17**. Name also starts with M: **Milutin-Mark**. Has Midas Touch. Shake his hand today. Amplifies the Echo.

Larissa as Amelia likes ketchup. As Pamelia, she called it catsup. She likes experimenting with her hair. She likes peace. Lyrical genius. Midas Touch. Always runs into someone famous. Wants to live off the grid.

Ethan Hawke who plays Tesla in the new Tesla movie is 49.

Hetepheres was 49.

Nikola Tesla was 49 on 4-4-1906.

Amelia as Pamela-Pamelia was 49 (1951-2001)

Pamela was born in 1951, Larissa was born 49 years later. Magic trick.

Sneferu is 48 and will be 49 when he figures out who he is (next July.)

Elon Musk is 49. 2551 - 1971 = 580

The 580 Freeway runs through the Bay Area.

Kamal el-Mallakh was born in October, died in October, 3 days before his birthdate and at the age of 69. 1987 and 13 years later, he was born again as new Khufu-Tesla. "OKT"

Using the number 2600 and subtracting years from it will determine a connection to that Dynasty. People born in the 1980's and beyond can use this calculation tool. 8 = 8, 2600 - 2020 = 580 = 13 = 4 (4th Dynasty Now)

Reading heiroglyphs happens in 2 directions, depending on the direction the symbols are pointing. You read from right to left if the symbols are facing right= READING BACKWARDS

There was an Earthquake in Tonopah, NV on 5-15-2020 that was 6.5 = the largest in 66 years. 2,200 aftershocks. Of those 2,200 quakes, 237 were above a magnitude

3.0, and 31 measured 4.0 or above. The three largest crested magnitude 5.0. The New Tesla's Birthdate = 5-16

All anything is about is numbers. Titanic = everyone starts talking about the numbers. 1989 Earthquake = everyone starts talking about the numbers. Anything = numbers, numbers, numbers. You see why we are here.

Eight is not a number, it's a sound. ATE = A + T = Bird & Echo = BIRD CALL!

Your eyes control your fingerprints.

Highly Electric Gods make the power go out, hack & crash all machines, Wifi, security systems, infiltrate your cars like Christine, you name it. Stay Puft Marshmallow Man Frequency = who knew? (Sorry in advance, this is the disclaimer, read my book) JUST KEEP US HAPPY!

Tesla published his autobiography in February 1919. I published my first book in February of 2020. **101 years later**.

I was writing my book in January 2020, **94 years** after Tesla was conducting his Queen Bee interview in January 1926.

All things that happen this year are an Echo of that year. This year equals 94 years later. Now we are in the Twilight Zone with our story.

NIKOLA TESLA
WIRELESS EXPERIMENT
4-4-1906

WE ARE ALL CONNECTED
1-25-2020
July 10 Blueprints
Nikola Tesla Tin Can Phone Wireless Experiment
Physical Results of April 4, 1906:

MAGNETS   FREQUENCY OF JULY 10

J   B   8   4-4-2   7-10

Planet Khufu repeats   Jose Altimira 4-4 (1824)   Pyramids of Giza (7)
Menkaure   Nikola Tesla 4-4 (65 & 1906)   29.9792 31.1342
Language   Lacey Welker 4-4 (1982)   Coordinates
Airwaves   299 9792 458
Passcode   Speed of Sun
Jennifer

                         Jonathan Huey 4-4 (2000)
                         Christopher Huey 4-4 (2000)

                         Brett Bonini 5-16 (2000)

                   Juan (Johnny) Esquivel 6-29 (2000)
                         Mark Erickson 9-4 (2000)

                         94 days into the year
                         94 years later
                         Tesla was 49
                   Frequency of repeating 2's & 4's
                   Coca Cola can: Nikola's Echo

MOULIN ROUGE OBSCENE

The Red Pyramid, also called the North Pyramid, is the largest of the three major Pyramids located at the Dahshur necropolis in Cairo, Egypt. Named for the rusty reddish hue of its red limestone stones, it is also the third largest Egyptian Pyramid, after those of Khufu and Khafra at Giza. It is also believed to be Egypt's first successful attempt at constructing a "true" smooth-sided Pyramid. Local residents refer to the Red Pyramid as el-heram el-watwaat, meaning the Bat Pyramid.

The Red Pyramid was not always red. It used to be cased with white Tura limestone, but only a few of these stones now remain at the Pyramid's base, at the corner. During the Middle Ages, much of the white Tura limestone was taken for buildings in Cairo, revealing the red limestone beneath.

The Red Pyramid was the third built by Old Kingdom Pharaoh Sneferu, and is located approximately one kilometer to the north of the Bent Pyramid. It is built at the same shallow 43 degree angle as the upper section of the Bent Pyramid, which gives it a noticeably squat appearance compared to other Egyptian Pyramids of comparable scale. Construction is believed to have begun during the thirtieth year of Sneferu's reign (2590 BCE.) Egyptologists disagree on the length of time it took to construct. Based on quarry marks found at various phases of construction, Rainer Stadelmann estimates the time of completion to be approximately 17 years while John Romer, based on this same graffiti, suggests it took only ten years and seven months to build.

Archaeologists speculate its design may be an outcome of engineering crises experienced during the construction of Sneferu's two earlier Pyramids. The first of these, the Pyramid at Meidum, collapsed in antiquity, while the second, the Bent Pyramid, had the angle of its inclination dramatically altered from 54 to 43 degrees part-way through construction.
Some archaeologists now believe that the Meidum Pyramid was the first attempt at building a smooth-sided Pyramid, and that it may have collapsed when construction of the Bent Pyramid was already well under way - and that the Pyramid may by then have already begun to show alarming signs of instability itself, as evident by the presence of large timber beams supporting its inner chambers. The outcome of this was the change in inclination of the Bent Pyramid, and the commencement of the later Red Pyramid at an inclination known to be less susceptible to catastrophic collapse.

Coordinates: 29 48 30 N, 31 12 21 E

Constructed: 2590 BCE

Type: True Pyramid

Material: Limestone

Height: 105 metres - 344 ft - 200 cubits

Base: 220 metres - 722 ft - 420 cubits

Volume: 1,694,000 cubic metres - 59,823,045 cu ft

Slope: 43, 40'

[Dr. Rainer Stadelmann (October 24, 1933 - January 2019) "OKT"
Was a German Egyptologist. He was considered an expert on the archaeology of the Giza Plateau.

John Lewis Romer (September 30, 1941) British Egyptologist, historian, and archaeologist.]

aka NORTH PYRAMID
Largest of the three major Pyramids located at the Dahshur necropolis
in Cairo, Egypt. The third Pyramid built by Sneferu.

KING SNEFERU

(snfr-w) "He has perfected me"

Was the founding pharaoh of the Fourth Dynasty of Egypt during the Old Kingdom. He built at least 3 Pyramids that survive to this day and introduced major innovations in the design and construction of Pyramids.

Hetepheres I was Sneferu's main wife and the mother of Khufu, the builder of the Great Pyramid on the Giza Plateau.

Reign: 24, 30, or 48 years - 2600 BC

Predecessor: Huni

Successor: Khufu

Consort: Hetepheres I

Children: Khufu, Ankhhaf, Kanefer, Nefermaat, Netjeraperef , Rahotep, Ranefer, Iynefer I, Hetepheres A, Nefertkau I, Nefertnesu, Meritites I, Henutsen

Mother: Meresankh I

Burial: Red Pyramid ?

Monuments: Meidum Pyramid, Bent Pyramid, Red Pyramid

"Beneficent" ruler = "to make beautiful"

The new Sneferu is Ryan Lindsay Silcocks. 7-5-1972
He is 48 now. 49 next year.

He will find his specific Patterns throughout this existence. The precise dimensions and intention will always Echo loudly.

MEIDUM PYRAMID

The Pyramid at Meidum is thought to be just the second Pyramid built after Djoser's and may have been originally built for Huni, the last pharaoh of the Third Dynasty, and continued by Sneferu. Because of its unusual appearance, the Pyramid is called el-heram el-kaddaab - (Pseudo Pyramid) in Egyptian Arabic.

Coordinates: 29 23 17 N, 31 09 25 E

Type: Step Pyramid

Height: 65 metres (213 ft) (ruined) would have been 91.65 (301 ft) or 175 cubits

Base: 144 metres (472 ft) or 275 cubits

BENT PYRAMID

The Bent Pyramid is an Ancient Egyptian Pyramid located at the royal necropolis of Dahshur, approximately 40 kilometers south of Cairo. A unique example of early Pyramid development in Egypt, this was the second Pyramid built by Sneferu.

Coordinates: 29 47 25 N, 31 12 33 E

Constructed: 2600 BCE
Type: Bent Pyramid
Material: Limestone

Height:
104.71 metres (344 ft; 200 cu)
47.04 metres (154 ft; 90 cu) beneath bend
57.67 metres (189 ft; 110 cu) above bend

Base:
189.43 metres (621 ft; 362 cu) at base
123.58 metres (405 ft; 236 cu) at bend

SHAPESHIFTING

The Pattern shapeshifts. Last name speaks first name of past life. First name contains previous last name sound.

NIKAURE
MACABRE
NINIBRE = BRETT BONINI, BROOKLYN BRIDGE
B = Foot = Walking
BB = Walking Excessively

NIKOLA TESLA
BONINI BRETT = NEE, TE

WINCHESTER
HETEPHERES
     HARAS
     SARAH

(Imhotep)
the one who comes in peace

kh   f   u

Khufu

i   mn   n   t   u   t   ankh

Tutankhamun

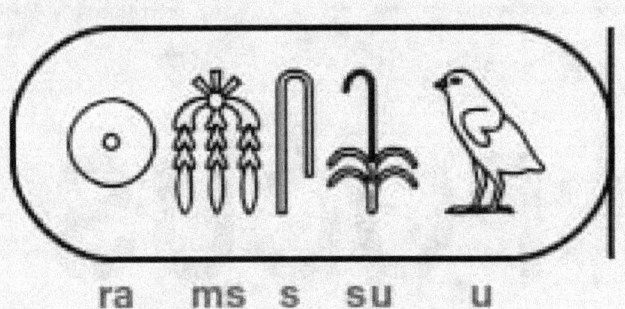

ra   ms s su   u

Ramesses

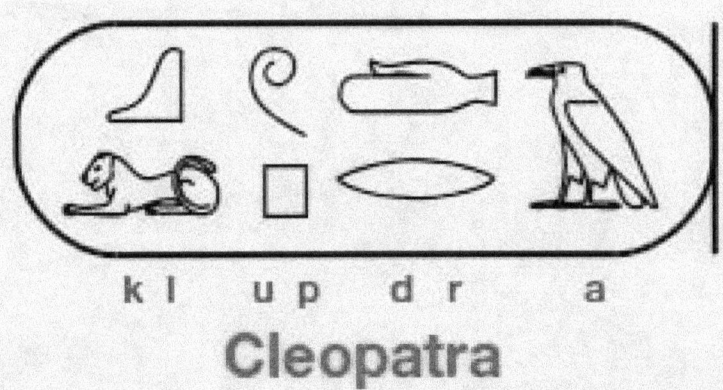

kl   up   dr   a

Cleopatra

So far, from 2600 BCE, I have found:

Khufu, Samuel Butler, Napoleon, Edgar Allan Poe, Nikola Tesla, Kamal el-Mallakh, Eazy E, Brett Bonini - 5-16-2000

Khafre, William Winchester, Wayne (Willie) Welker - 11-3-1949

Menkaure, Einstein, Brian Gene (Breinstein) Paulas Jr. - 3-15-2016 - 369

Khamerernebty, Elizabeth Blood Welker, Carmelita Isabel Esquivel - 9-21-1998

Meritites, Babie Annie Winchester, Amelia Earhart, Pamela-Pamelia Florine McGovern aka Tumbleweed, Larissa Elizabeth Esquivel - 10-4-2001

Hetepheres, Sarah Lockwood Denise Pardee Winchester, Natalie Wood aka Natalia Nikolaevna Zakharenko, Lacey Lee Welker Paulas aka Genie Greentrees - 7-10-1980

Sneferu, Ryan Lindsay Silcocks aka RLS - 7-5-1972

Djoser, Jose Altimira, John Hansen, Juan (Johnny) Esquivel - 6-29-2000

Djedefre, Jesus, Jason of Green on the Go - 4-3-1985

Found but not yet connected to the beginning:

Vincent Van Gogh, Brian Gene Paulas - 10-23-1990

Harry Houdini, Farooq (Frank) Helmand - 4-20-1983

Selena Quintanilla Perez, Selena Gomez - 7-22-1992

Mark Twain, Samuel Clemens, Mark Erickson - 9-4-2000

Mystery Twin Brothers: Christoper & Jonathan Huey - 4-4-2000 Connected to Tesla's 1906 Experiment Picture 4-4-65/1906 Labeled "5" & "6"

Ages of those in my tribe: 4, 18-19, 20-20-20-20-20, 22, 40

The house I live at is Frequency of Edgar Allan Poe and Charles Lindbergh. It's a talking ship, a giant Vessel. This is how weird Science is: Infinity. The story has always been the same, just different. High Frequency will show up as High Frequency in any Pattern. The proof is in the pudding. Doing things repetitively to keep things exact helps. That is how I've gotten the whole story to play out in order for me to see it. I had to make the snail dance out of its shell to dissect what it is. How it operates. It's not hard, it's just tedious and takes endless patience. And that's what this has all been about. Thankfully this is good news and there is hope after all. We can get back to being the Gods we are.

Anyone who appears to "die young" was only having a dream within this dream. Ask Edgar Allan Poe. This is Fairyland and all his poems.

EIGHTH WONDER OF THE WORLD
AYT = 8 = Eight
A = BIRD
T = ECHO
2600 = 8
PASSCODE = 8

2600 Then	1980 Lacey
2020 Now	1922 Sarah
580	58

SLITHERING

How the Pattern approaches you. The land is a serpent dropping mice on your doorstep.

To catch a Tesla, I had to stay home and limit my contact with those I see and meet so I could attract my specific crew. Harnessing my Frequency. Containing it like fireflies in a jar. Bricks are laid to step on and they are installed with the Pattern. Each shoe, foot, footprint is specific to yours. It will come to you if you sit still and let the Planet.

2017: I met Italian Joseph. I met mysterious Osiris (Juan), I met Egyptian Farooq (Frank) Helmand in June, then James Woods. Car guys.

On 6-16-2017 (39 years after Grease was released), I had an encounter with their 3 black Stingrays. I met all 3 of them that night under extremely different circumstances. The 3 began and so did this show. I was 36 turning 37, 3 weeks later. I spent 7-10, my birthday, with Frank. (Imhotep) I also spent 7-30 with him on the beach in HMB with a fire, smores, RED Gatorade, jerky, nuts, and a fire. The Sceptre being born. People should ask Frank Helmand about his Sceptre that got us all here today. It's attached to our LEFT side.

2019: I met new Tesla in October

J is the land. Land hands. Land Echoes. Chariots. James Woods is the land Echo of Edgar Allan - Stay Puft Marshmallow Man footprints coming: Since Tesla was Edgar, James came to let me know the Echo of Edgar was on the way.

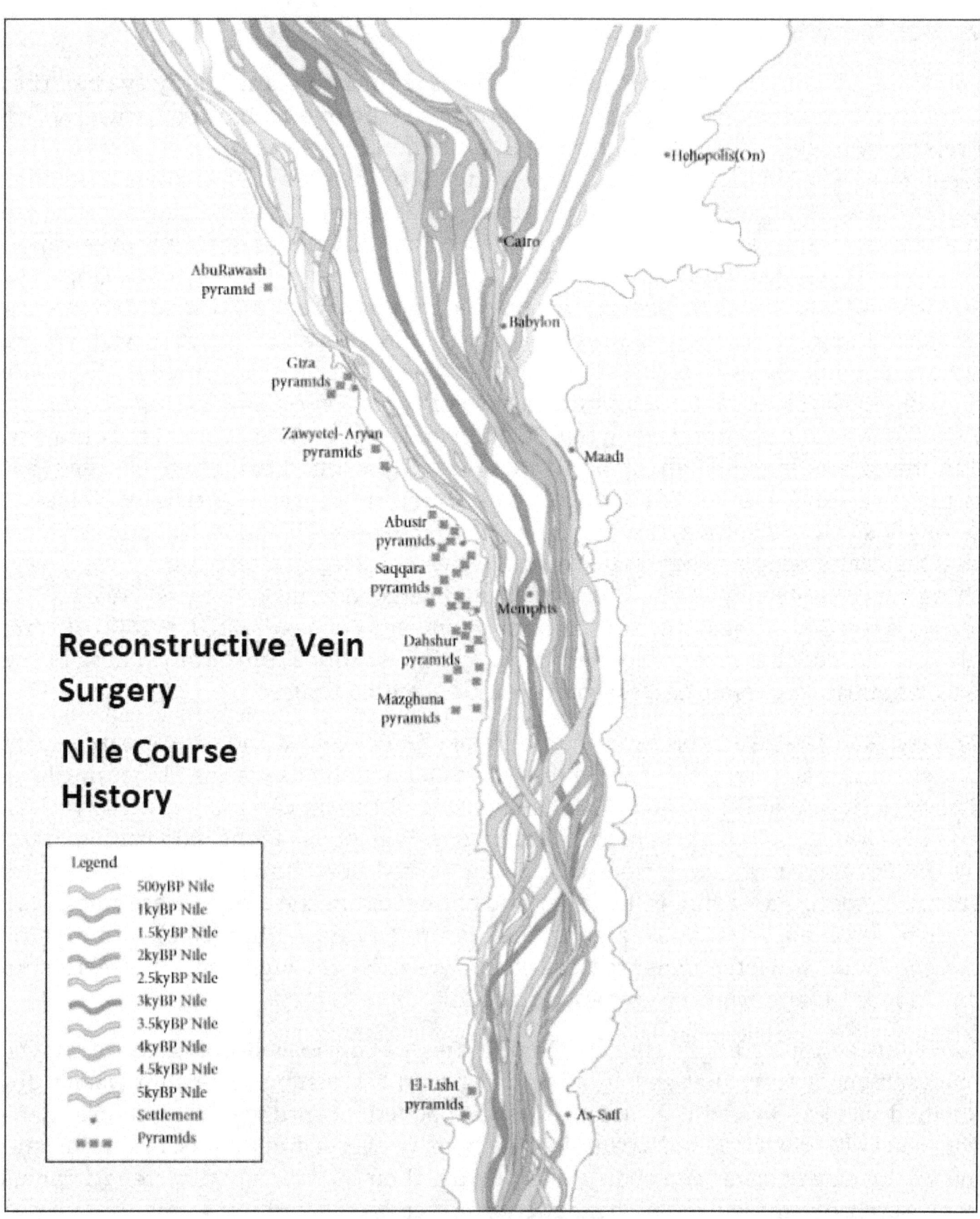

Reconstructive Vein Surgery

Nile Course History

Legend

〜	500yBP Nile
〜	1kyBP Nile
〜	1.5kyBP Nile
〜	2kyBP Nile
〜	2.5kyBP Nile
〜	3kyBP Nile
〜	3.5kyBP Nile
〜	4kyBP Nile
〜	4.5kyBP Nile
〜	5kyBP Nile
*	Settlement
▣▣▣	Pyramids

*Heliopolis(On)

*Cairo

*Babylon

AbuRawash pyramid ▣

Giza pyramids

Zawyetel-Aryan pyramids

*Maadi

Abusir pyramids

Saqqara pyramids

Memphis

Dahshur pyramids

Mazghuna pyramids

El-Lisht pyramids

*As-Saff

9-26-2020 BACKWARDS

You think you are on "Planet Earth", but this is actually a giant brainy eye ball that conducts its business through trees. The trees are the ones catching themselves on fire, communicating to create lightning that catches themselves as structures on fire. Firestarter, The Movie. They contain the Patterns of the past in their rings. The rings chime out symphonies in waves. The Patterns that appear as everything physical are the waves. Every day, everything contains an Echo. Each one of us are in our own Echo chambers. Mumbling out a bunch of chimes. It's not words, it's not talking....it's Storytime. Just a bunch of land characters singing the same thing with different variations of Frequencies, in the same coordinates, following a trail there as well. We are square dancing. One side is this side, the other side is the dark imprinted shadow side dancing in the same motion, off by a Pattern of beats. Everything is sing and dance. Just like the Echoes imprinted in our Calcium and Copper. The Echoes are contained and travel. Singing and dancing is all we ever really wanted to do here because that is the first story; our foundation. Our true Founding Father. Electrically etched as strongly as all the heiroglyphs on the walls that we dream about. Nobody can take that away and people need to understand we were asleep in Utero until now. Everything was remote viewing before, also strongly etched while still being created. But we were the Gods that all the stories were about. Our Echoes created in 2600 BCE are playing out now because we are here now. Again, but more...physically. Think of the way we started as Venus has, gaseous dreams. Sleeping dragon.

The results of Tesla's 1906 experiment are playing out now and have created exact target coordinates because of the trees; groves planted by us via the Planet in Utero. The background of his picture: TREES. The Echo of the tin can phone, string (trees), and everyone's contribution to the experiment. This house is the wood ship parked by the Eucalyptus grove of trees that Sarah parked her Titanic houseboat next to. Different coordinates, but still similar and connected in the story. These trees were planted as an experiment as large as the Pyramids, Khufu Ship (tree), and organic matter found in Hetepheres treasure tomb. We exist through these Vessels....ship and blood. Targets acquired - magnetic circuitry.

Trees buried under bricks ringing-singing dreams to those sleeping above. Maybe it wasn't the Sphinx that gave that sleeping man the visions. The ship wasn't uncovered until 1954 and that dream story happened a century or two before that - that's how I know wood talks. You sleep next, near, over it and it talks. Dreaming and awake. The trees have been ebbing and flowing their electric songs, magnetizing us back. Sands of time = glass. Also magnetic, refracting.

A target of ginormous proportion was acquired by an experiment. The surroundings of these specific numerical coordinates were intercepted by an electrical impulse = The Death Ray: Tesla's Intention. The voices and actions recorded by the trees as well as the snap of the picture were the nuclear bomb that created this wave of mutants: US. Tesla's intention created everyone in his Echo. Everyone now has proof. They can trace themselves and see their magnets. He made things precise. Patience was all that was required for this experiment. It took 114 years to create this Dynasty where we could confirm and always remember what we are working for. Now we can do what we want. We aren't going to have to worry anymore because eventually, the Echo engulfs the entire Planet. He did what a laser eye surgeon does. It doesn't just stop at Tesla cars....it's Infinite. TKO.

We will all be walking around with glasses instead of masks and we will be nice geniuses again. We will sit down with the trees and each other to eat our peanut butter sandwiches again.

Everything is a call from our past life. Thank you, trees!

PIONEER, CA

Intergalactic camping in 1993-94: Waking up to see the sky light up in the middle of the night. Then having my head fondled in the night on another trip.

2020: Hearing the story that there were recent recordings of a pair of mountain lions in one shot, then one staring at the creek alone. A bear and her cub were also spotted. I was sleeping on the ground when I saw the light. When I was fondled, I was sleeping in the VW camper with the back open. I cried and Khafre had to crawl in and sleep with me. What else is there that we can't see? Learn through stories of others.

Person recording animals: Name starts with J

GUND TEDDY BEAR COMPANY 1898
I found a copy of this bear sitting beneath trees in 1985.
Ursa Major - The Great Bear - Constellation

SOLVING THE PUZZLE

Pamelia's actual birthdate may be 1951 as I was studying her baptism certificate. Google says 1951 as well. My birth certificate lists 1952. My baptism certificate from 4-4-1982 has her name misspelled as well. She was 49. Hetepheres was 49.

6   7

Pamela Florine

9-28-1951   10-4-2001

8   8

49 = Pamela was 49, Larissa born 49 years later

7   8   7   8

Pamelia McGovern   to   Larissa Esquivel

2-20-2001   10-4-2001

7   8

Connected by **I** & **1** (Babie Annie)

Everything is connected by multiples of 4 in simplest form: 13, 22, 31, 40, 49, 58, 67, 76, 85, 94

Add & subtract days between dates, years between years to find the Planet's Pattern. Write it all out.

All break down to 4 when added, until you can add no more and the answer equals 4.

Strong land Pattern exists and shows itself within the Pattern of 4.

4th Dynasty = Strong land Patterns (Follow the letters shapeshifting into new people-characters)

◆ ◆ ◆

Best Ancient Games: Memory, Boggle, Scrabble, Chinese Checkers, Dominoes, 4 Square, Simon, Simon Says, Tetherball, Red Rover, Cards, Soccer, Baseball, Hopscotch, Balance Beam, Double Dutch, Hula Hoop, Charades, Rock-Paper-Scissors, Shadow Wall Animals, Hide & Seek, Marco Polo, Hangman, Guess Who, Connect 4, I Spy, Counting Everything, Writing It Down, Dancing, Singing, Chanting, Experimenting, Building, Digging, Storytelling, Show & Tell, Word Search, Puzzles, Jacks, Being The Planet

9-23-2020

Einstein and I went to the store today and of course, when we come out, the Mercedes from the Blinding Lights song was in the parking lot. It was quite amazing and I couldn't stop staring. We went to the park and saw the "TESLA IS ALIVE" I wrote on the sidewalk when we found someone had left their leftover chalk. It's been there for almost a week and other little kids started adding their decorations around it. This is what happens when I go out. I run into magic because I perform magic everywhere I go. I put intention into making each scene a stage just by having fun with it. I don't sit or stand around bored; I collect tree trimmings and make pictures for others to see. I like messages in a bottle happening to me, so I do them.

The total at the store was 11:44.

EXPERIMENT ORANGE:

I poked 3 oranges on this bare tree. The next day, neighbors are harvesting their citrus and give me 18 oranges.

31

9-22-2020 LIVE HERE & THERE

That's where communication comes from. You get to download a rainbow of information just by existing in different coordinates. Living Infinite lives is the idea. We should be nothing but travelers.

9-21-2020 THE MAGICIAN'S TRIBE

I'm not crazy - I"m high Frequency. Those with more blood will understand this. If I give Jovy Chow a copy of my book, she will remind me that I'm not crazy because she worked with me for years and we've known each other for years. It's not a coincidence that her son drives a grey Tesla. It's also not surprising that her name is 4 4 and starts with a J. They also have a dog, smile a lot, and are very friendly.

Tracy Bonal lives across the street with David Bonal, both Edgar Allan Frequency and their address equals 4. Tracy rhymes with Lacey and her daughter's name is the Frequency of Nikola Tesla (Lauren Bonal.) They have a pool and a Hawaiian theme. Also, an elevated white house with a palm tree. We met at Remax Accord where I was processing her paperwork to start as a new agent and I saw her address was the same as mine with one number off, but directly across the street.

We can't forget Harry and Louise Metaxas who moved down the hill from me, literally 5 or 6 houses away. They were my previous landlords for at least 8 years before I moved here. They previously lived on the border of Oakland, then moved here about 2 years after I did and I've been here for almost 9 years. Harry is Greek and Louise is Egyptian. What a perfect match. Romeo and Juliet. She drove a 2 door silver Mercedes with their name "METAXAS" on the plate. He still drives a white Ford Explorer. That was 4664 Heyer Avenue, Unit D, the only unit upstairs with wooden fencing enclosing the porch. There were 4 units. The other ones were upstairs/downstairs and ours was in the back, at the top of the hill, flying in the air with the Khufu Ship fence. My neighbors were Greek who moved in a few years after I did. They had Greek statues painted on the walls inside their house. It was breathtaking. They also gave us a lot of meat and Baklava. They were the nicest people on the Planet. That is what our future looks like, minus the meat. You know I went school with their son, Jim Psarakis. I do believe we were in kindergarten together before I moved away to Florida. So he's a perfect example of an old school and high Frequency reminder. J's are everywhere. Genies.

Up the hill from this place, there was a younger guy named Alan who drove a black 2 door Dodge Challenger. He went to school in a different town with someone I used to know who lived here for a while.

On the other side of Echo hill in the red house is Julianne with the killer dogs who can't stop building onto her large property. It has come a long way. One time they left it painted pink for 2 days before they painted it brick red again. I met Joseph there one night along with his Cannabis toting friend. That was during one of my Extra-Terrestrial excursions. They were very nice and they also knew a "Frank with all the cars" just like me. Lots of Motorcycles go there and they have a tree trimming business.

Both associated with trees. High Frequency. I was told to go there one night, don't ask me how, but I came across the mirror, the looking glass. What I am looking at is also looking back at me. So the gong is across the way and all around this rim of a basin town. Castro Valley is Tesla Valley and this place is sacred Coppertown beyond belief. This place GLOWS GREEN.

People have to know what the Bay Area is. What water does, especially flowing water. It empowers us just by breathing it in our lungs, letting it fall on our skin creates waves of Echoes to wake you up. This is Mecca. This is where it all began, even Mount Diablo is labeled the entrance to the Planet.

I was sent a picture of it on fire from my friend's back yard. His name was Nick DeNijs. He sent me a video of him shooting a bow and arrow in his back yard as well. He looks like something you'd see at a powwow. And that's how he hangs out. In 4th or 5th grade, we reached into a jar to pull a subject to present a speech about. Of course I chose Mount Diablo. I had to memorize-repeat this long thing and I remember being so nervous about it. I was like, I just moved back, why do I have to do this about this land I didn't feel part of anymore? I just remember being reassured that it would be great, I memorized it easily and then presented to an entire cafeteria, powwow style. I think there were feathers present as well. We are definitely connected to the story embedded in the dirt if we are associated with it. This place is so, so special. I'm so grateful.

ECHO X-RAY

I have Larissa's X-Rays on the counter from 10-1-2009. 3 days before her 9th birthday, so she was 8. She fell off the monkey bars and fractured her right forearm. Right side of the body is the mission now. Connect to the Statue of Liberty. Amelia was associated with New York, planes, and structures. Her pictures prove it. Osama was as elusive as Amelia. Once they found Osama, the Planet already had me penciled in advance, knowing I'd pick up and figure this out. So the story played out. The Planet programs this timed stage. Your wishes are of the Planet's and you both join forces to do what needs to be done to restore harmony here. When you understand people don't die, this place turns into Heaven. The Heaven and Hell stories we've been hearing about are of this place. There's no raging pit of fire to burn us apart from the Sun, but in dreams, we exist through the Sun, with no body. Visions are produced, but there is no pain apart from high Frequency, and you get to be a part of something huge. Remembering your dreams shows you what you did while you were in the Utero dream. Your contribution.

Larissa chose an orange cast for her right arm. The Statue of Liberty is made of Copper and she holds a torch at the top of her raised right arm. Since the color she chose was a mystery to me back then, I am able to connect it now, 11 years later. 10-1-2009 = 4

The Statue of Liberty holds fire to the sky. Amelia Earhart was the messenger of 9/11. "Fire 2 the sky"

Used for this year's tribute were 2 blue lights 9-11-2020.....Larissa also associates "19" because she is 19. It's been 19 years. The Statue of Liberty is now blue. What happened to Larissa's arm is the Frequency of the Statue of Liberty. Fractured. You can see the curse now. Thank you Edgar Allan Poe.

From: Amelia 1939
To: Pamela 1952
                13

From: Pamela 1952
To: Larissa 2001
                49

Hetepheres was also 49 and associated with 4.

Meritites 2600, Annie 1866, Amelia 1897, Pamela/Pamelia 1952, Larissa 10-4-2001, all associated with the Pattern of 1 - I. 1 looks like I

Each one has to do with repeating Patterns and being connected to 8.

Pamela turned to Frequency of Pamelia on 4-4-1982

She came back in "OKT"

New York,[1] U.S.

Coordinates	40°41'21"N 74°2'40"W
Height	Height of copper statue (to torch): 151 feet 1 inch (46 meters) From ground level to torch: 305 feet 1 inch (93 meters)
Dedicated	October 28, 1886
Restored	1938, 1984–1986, 2011–2012
Sculptor	Frédéric Auguste Bartholdi
Visitors	3.2 million (in 2009)[2]
Governing body	U.S. National Park Service
Website	Statue of Liberty National Monument ↗

◁ ○ □

THE WOOD CHAIR

Babies and children are not babies and children. They are pure Planet Frequency reminders of the story and are also involved in the story of the Planet within your Blood Brother Tribe. They bring messages. When Einstein repeatedly brings the old wood chair into the kitchen, that's the Planet doing that for the story that's under the kitchen. The story exudes through our surroundings, the bricks, the tiles, the wood slats, the glass windows. It's a giant screaming Medusa head. It's not just a house, it's not just dirt. There are stories burned and buried in the coordinates. Patterns dating back to no end perhaps. So that's how you know this is a song with Patterns you can change. Each area is playing a certain song. You have to change the song by dancing and singing in its place. That's what Einstein has been teaching me to do: Sing & Dance. It's really nice to be reminded to do this. That's why I love having little ones around because they still live the story and they send direct hits to your Battleship from your past life. He just kept bringing the chair, bringing the chair. I was guided to figure out Hetepheres was me on 9-5-2020. That's pretty funny as I've been working on this since 2017. I'm glad that knowledge falls into place every day like the child's room simulation in Monster's Inc. I feel happy knowing that each action creates a ledge for a reaction to sit on. I love knowing this place Echoes and I love capitalizing it, too. I just wish for everyone to know the magic of this place. I just want to feel they understand what I'm trying to say. I repeat the same thing over and over every day, like a bird, but I don't care, because it's all that I can think about and it's all that talks to me. I'm grateful. Thanks Planet for not letting this be for nothing. I hope that makes you smile as well as everyone else here. I'm tired of chiseling the same message but I have no choice until there's movement. So each new wake up shows me a new version of the day before. Groundhog Day. There I have placed all of my surprise eggs. The Planet rearranges them all for you and you know they will be there because you live a life of Pattern. Why not? What else is there to do here besides GONG? Someone will believe all this madness is me sitting here casting lightning bolts from my windows one day. They haven't read my notes yet and I haven't included even a portion of what's written so far. Each page of each book is only a giant one eyed monster waking up and morphing into a new character each day. That's all this is....me, growing into a giant Serpent, undetected. Charlie's Angels.

Khafre mysteriously gave me $100 today and told me to take it. I was thinking of him as Edward from Pretty Woman yesterday as it was just him and I here at the house for the last 3 days and I've done nothing but stay in my hole and listen to the sanding and banging of his painting ladder. He's in the process of fixing this Sphinx structure. He said the money was for all the electricity he was using.

GREENTREES 1982

9-20-2020 THE LAND IS AN ELECTRICAL SERPENT

When you speak into the glass or any echoing thing, it pings back into your Frequency and that's where the Echo is created. By creating one. It will ping back into your Frequency circle = the ones connected to your Frequency. They will experience what you enhance. Smoking amplifies the response by expanding the Frequency, clearing it out for the true story to attract. Becoming the Frequency of a tree is becoming the child again, where all of the magic communication is happening through what we call dreams. It's just our Echoing mall service. Once people learn about what this Planet actually does, they will change their ways. The wave of the hand is a wand waving vapors through the sky. We create the clouds, we create everything with what we do with our hands, how we think with our hands. We don't realize our hands and feet are also brains connected to a larger brain. We are Starfish, each limb is magical and means something as sacred as any animal or insect you see. They choose to show themselves to us because they were told we were the ones who have been writing the story. Most people just fan animals and insects away because they were told to associate them as different, not something to talk to. Same goes for trees and structures. You are actually supposed to be talking to the walls of everything, saying hello and thank you. Can you imagine walking down the street and seeing people talking to walls and then smiling at each other? Imagine that is our society. To create a wave of gratefulness only takes one other's acknowledgement. That's how extreme the Frequency of Copper and Calcium is. Skeletons glowing, dancing in the dark. The battery we are.

Nikola Tesla was connected to July and January, both J sound. Edgar Allan Poe was connected to January and October.

9-19-2020 PASSCODE "OKT"

Everything is programmed with intention. Each thing is its own function functioning among all of the other functioners. Everything is tied into this Infinite brain system. Even the Sun rising every day is showing us a system. A system is running. All the Planets are components and sway together in song. Everything is following a beat. Everything is timed in perfect systematic format. It's genius compared to the time system that we have been running. Planet counts in tally marks. Music notes. If a number exists because of any calculation, it's already known and counted by the Planet. So 114 also equals 6.

I had a dream vision of the "OKT" passcode and face. The secret knock of this place. Fright Night has the correct interpretation of that vision. An abnormally large mouth screeching OKT was the imprint and I was looking to my left like most of my dreams. And the process was showing me a high pitched type of encoding, like screaming as loud as you can as a shadow face, but showing the message of October. It contains 7, 8, 1, 10. It's the Magic 8 ball of sounds and words. All the communications having to do with this sound are high Frequency. All signs point to October. Even Spanish Octubre has it right. The reminder that it has been starting up again, coming back online. August is also the 8th month. That has everything to do with the story as well.

Everytime you hear the "T" sound, that's Khufu. That's Tesla. Khufu was Khnum-Khufu. Edgar Poe turned to Edgar Allan when he added it. There is a Pattern in everything. It exists in every single word. In each word is a world of sounds and meanings. Just by a word existing, there is another world that has had an Echo world created for it. We all have access to all those worlds, and together. Call them stars, call it an invisible world that you can only gain access to by understanding the Planet codes. This place contains the answers all around. The Nickelodeon exists all around. Each thing you encounter is a poem. All the sounds can be traced back to the beginning. Beginning of recording is the true beginning of time. Like now. This is just that starting all over again. If you understand all the stories, then you can begin to connect the dots like the stars stringing the story along in another fashion. It's just a new wave, but it's the same story. Now it's really the same story and we have a heartbeat on the Venus fetal monitor to prove Echoes are created and everything is heard, even silence. Everything is being created as we speak each word. You think someone is watching you but it's actually just the typewriter clicking. The Khufu story and sounds are real. Just when you think things have ended, they have actually only just begun. It's okay to understand that the future is actually just the Ancient returning. The true geniuses would exist and come again because of some great experiment that happened to occur in nothing but dreams. Everything has just been directly

Echoing our Return. And here we are. Here we will come again. When you understand you will see your friends again someday, it will make you feel relieved and able to WD-40 your way through your important tasks. I couldn't move or do anything for hours this morning, I was just standing in place staring at everything. I could only think that I didn't want to do anything. Then headphones popped into my head and music that I recorded from a year ago from the FM radio. Those are the Echoes that are happening now. I titled the songs then and now I see the Patterns playing out like Broadway around me. Like the Blinding Lights song. I then came into the bathroom to smoke, type and listen to music. Gods need to be excused to correct the Planet. Gods living like they are not Gods are haunted Gods. I am just here to write the story. If I write more, more things will change. So this is my day. Recording 2600 BCE happening is a beautiful Macabre. The triangle Sun shining down is what we've become. Eagle Eye. Choose or perish as Gozer says. You can choose to perish or not. Understanding is the choice.

Dreams are symbols-chapters that represent the Pattern that is coming in to surround you. It will make sense after it all happens. Each infusion of Frequency will stay at the top of your existence as a title to each day. The dream is the title that hangs over each day. Your most recent dreams are the most recent chapters. Dynasties popping up as little gifts that you were promised would be there when you asked. Each dream is a phone booth. Call the 900 number to your heart's content. It might be prerecorded, but it's worth discovering and experiencing.

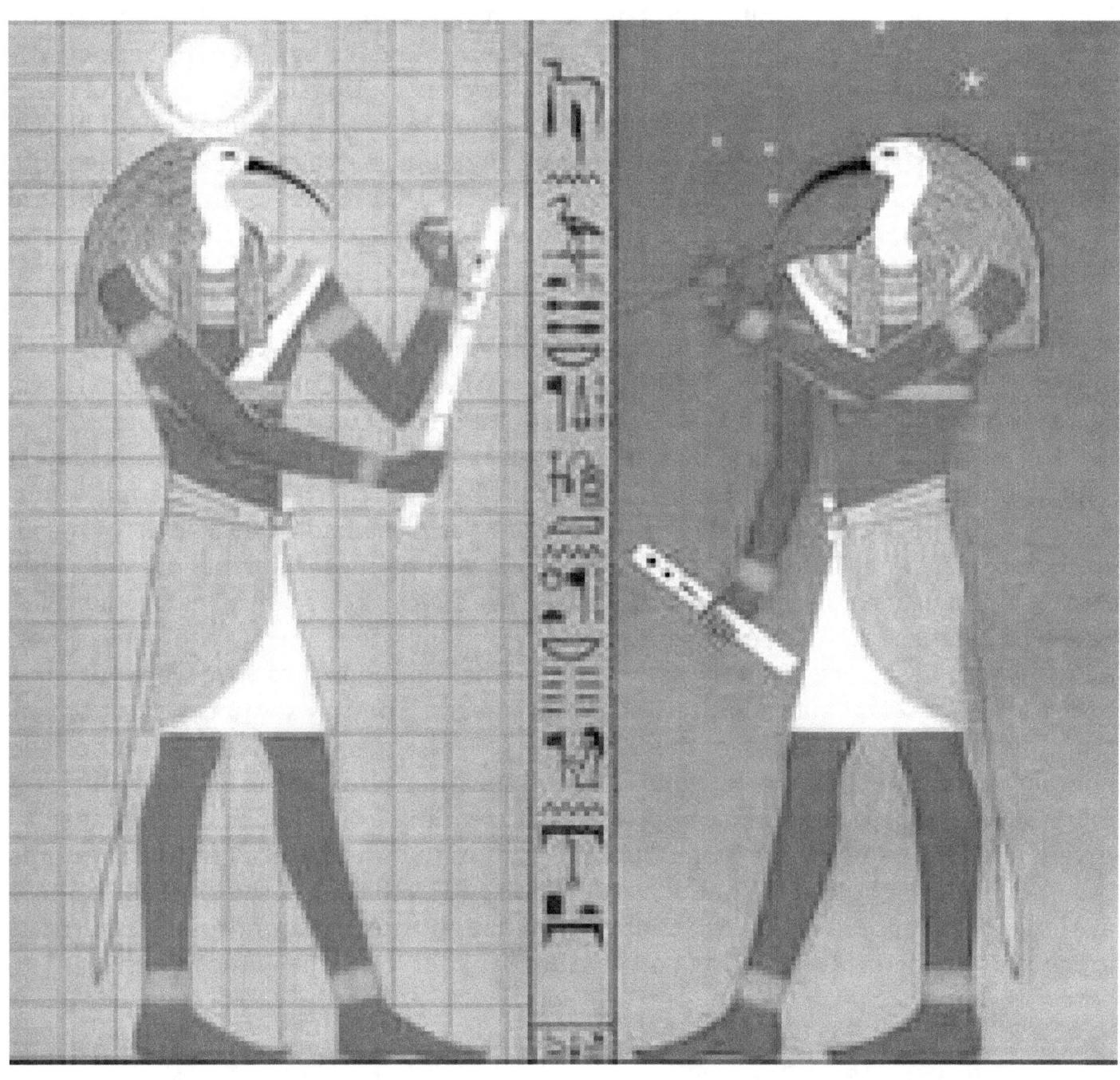

9-18-2020 WATCH WHAT HAPPENS NEXT

It's 8:15pm. About 10 minutes ago I asked for a sign and then a shadow dog with a blinking light collar arrives at the crosswalk below. I hear dogs start to bark everywhere and then it hits me: They are playing Marco Polo. Dogs everywhere. As I was standing at the window, a dog starts barking at me from the neighbor's yard below, but it was a visiting dog. It heard me talking my dog language and acknowledging Marco Polo. That's when I heard the bark along with all the other chiming barks here on Sesame Street. This is Barkley's World. Something about dogs. Ancient Egypt and dogs is where I am told to go next.

Einstein and I went on a giant walk in the morning. We walked to the upper dog park and then to the lower fun park. We played and only a handful of people were there. It was more enjoyable without all the noise. Someone left us chalk so I wrote "TESLA IS ALIVE" in blue on the sidewalk next to a rainbow that the original friends drew. I kept saying we needed to get some chalk. I can't believe I didn't make a "Hop Skotch" with it. That's why I was wanting it. I am sure the trees had their stories in mind, so thank you Planet Trees. We also observed 4 green paint handprints on one of the Eucalyptus trees inside this little fenced area I've always gone since I can remember. I used to balance on the tree fence. So this is where Einstein and I go now. We sat on a Eucalyptus yesterday together, so I wonder what happens next. We were out for 4 hours and saw 2 red and 1 white Tesla on the walk. Then as soon as we got to our front door, the red Tesla passes by. "Highway to Heaven" would create an Echo. The characters exist as the electric snake land and the mumbling of the Planet = John and Mark, 4 + 4 = These are all just blips on a neverending graph radar that we need to invent next. The sounds are just a tracking system to be accessed. Tracking systems invented in Ancient time. What does the land send? What does the sky send? There are lowers and uppers. They each connect and have hunter view or drone view. Those are worlds in themselves. Saying you are thankful out loud, the Planet listens.

9-17-2020

Four is the only number spelled with 4 letters as symbols to represent a correct Pattern of Frequency. Four is the only word we were waiting for. Four is the only word that makes sense. We had to wait to invent "English, The Complicated" in order to figure this out. Maybe it's the 4th Star. I wonder who is on Venus if we exist through water and other vapors. Did we create a new way of our existing through all of our experimenting? It's a heartbeat of something miraculous. Maybe it's a pinhole, a needle.

one two three

FOUR ✓

five six seven eight nine

9-16-2020 GREEN GLASSES

Firefighters are the Ghostbusters. This is the part where the sky clears and Dana is standing on top of a car with Peter Venkman and they have marshmallow in their hair. We ate marshmallows yesterday. Everyone sees they are the Superheroes. Rick Moranis as the new Tesla is standing around asking if anyone wants to interview him. "He was an eye witness." Nobody listens, they just want to get him to the hospital.

The Black Cat in the back yard this morning acted as if he owned it. Crows in the front yard for peanut breakfast.

Life is an Edgar Allan Poem and I'm living it. Thanks, Eddie, Edward, Eddie Money.

I have to stay high all the time until someone acknowledges what is happening. I have to stay high to accept this reality. To remember it's not real and to remember my truth and the true story. What's going on here is not something I want to be present for, except for an extended experiment. And that's what I remember being here for. 4-5 years in a Cannabis induced existence and then writing it all down sounds pretty special...and it has been. It has been both a challenge and a gift. But the Planet has shown me that's what this plant is for. To sit and remember to observe truth and then do it. Really do it like there is nothing else to do but save the world by surfing from your couch. Trust me, it does something!

This is my mission and now that it has been made readily available to me to complete, I'm doing it. No matter what anyone says. Because that was the main wall - "You don't HAVE to do anything." I have stepped on the ship now and steering it with what I came here to do. The plans tell the ship where to go. That's just another topping on the sundae that tells us that it is okay to smoke = it's existence. Treated like a slave and smuggled like gold? Come on, that can't go on for long apart from only being part of this story. It all happened for me to write about in my world. What I encounter, what the Planet presents to me is specific. Because of this plant, I am finally shown the Pattern that has been eminating through these past 40 years. And it's happening throughout our entire existence.

So what if it makes everything specific and you are shown the correct vision of the child Frequency known as original Planet? It makes you see clearly and it lets you see your story. It lets you become the Hunter again to find it. Yes, that is something that has been obstructing our view: Them.

There is no way to put prices on things. The Planet will revert to a Bucking Bronco if you try putting your muzzling words-sounds on the faces of its creations. Understand the calculations you are speaking. Being told to repeat everything you hear and see is keeping birds stuck in cages. You can't make Gods do ANYTHING!

9-15-2020 "A CLUMP OF CELLS" ON VENUS: THE HEARTBEAT

The electric company is here at 10am. They are working on the gas meter. Brooke, the technician, told me it would take 15 minutes. This morning I see a video on Youtube that they've found life on Venus, gas life. Last night when Einstein came home, he was farting everywhere. They were really stinky and it's rare when that happens. Today is 343 egg day from 1911.

A couple of days ago, I watched a video of a walk across the Brooklyn Bridge in New York. I watched at least 5 minutes of it. I felt like I was doing it and then my left side tingles. The blood has begun pumping like Bella the Vampire opening her new red eyes.

The gas meter was changed in 56 minutes. And the BR Echo was here.

CB RADIO

The wood is what we use to get here. Communicating with/through the trees. We are the Frequency of the trees. We are the same Frequency as the trees. If trees are a radio station, we are each a radio station. Trees bring interference. We come THROUGH the trees. When we smoke Cannabis, we become the pure Frequency of the trees again. The "child." New again, free, original. It's pure like Reverse Osmosis.

If you become the trees, the story surrounds you the more you never leave. Then you are stuck in a mall with no people in it. Echoing by yourself, the clearest connection ever and 1 person keeps trying to join the chat but the connection is dial up. So you just wait in this Echoing hall room, waiting for someone to come online. It's the 1990's. Like I said yesterday, it's 1906. Maybe it's also 1996. Imagine the vastness of all the dates of things that happened connected to Tesla and his 1906. All of the dates between his invention patents to now. All the dates in between those dates are all Patterns of the Planet communication. Just subtract years and break those down to simplest form. "Between the lines" = extracting the caterpillar from the bud. "It's so hard" = Yeah, because no one has never done it. But that's the heart.

9-14-2020 REPLACEMENT OF FREQUENCY - A SACRIFICE

After the 1989 Earthquake, a month later, I moved back to the Bay Area from Florida. About 1990 or 1991, my elementary, along with others in the district held a gigantic choir powwow at Centennial Hall in Hayward. It is at the center of the fault, was damaged and has been closed since. It hasn't been torn down as of yet and has just been sitting, but we haven't had any big earthquakes. How interesting that it's a giant Echoing building that was full of children singing the "Echo Song" as loud as they could to each other. We were a bunch of schools singing back and forth to each other like Red Rover. That's coordinates singing to other coordinates by familiar Frequencies in a different coordinate. Heart surgery, a stint. Our voices were sounds that became glue to fill in the cracks. That building blares the Echo Song to let everything and everyone know that we are here. And that's okay.

Around that time, I was assigned to play the middle part of a 3 part glowing skeleton in a black light graveyard in Tom Sawyer. The focal point of the story is that the eyes were looking at what the other skeletons were doing, and eyes are white, so they were obviously glowing. Sphinx and Calcium Echo. There were 3 sets of 3 part skeletons. So 9 sets of eyes looking left to right. All I can envision is 3 sparkling Pyramids. Tom Sawyer is Mark Twain's Echo. All of this was orchestrated, you see? It was arranged by the coordinates, the trees. I just came back and had all these things to do and I didn't even know it. That's how it works. We are blinded, forget, then have to learn to remember it all again. Just as we forget our dreams. Each time we wake up, we are waking up from traveling through the Sun. Electricity zaps it away as it reminds us of what we are traveling here through and as. The blinding reminds us that this place doesn't belong to us and neither does any of this "stuff." Each symbolic thing we put in our hands is part of one or Infinite experiments. Nothing belongs to anyone. Nothing here has belonging; that's just a word. Anything is only for helping us to remember. Gods don't own things.

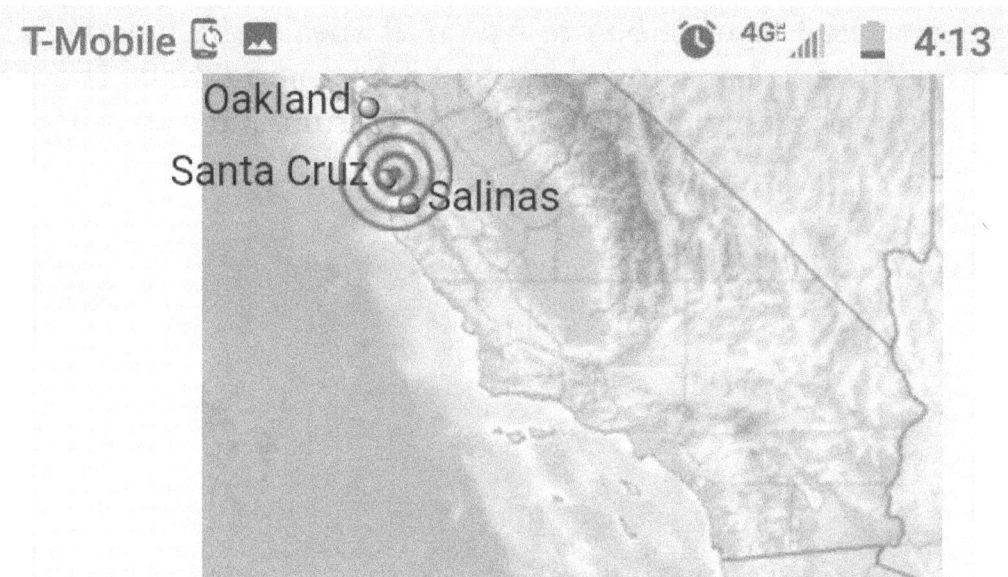

UTC time	1989-10-18 00:04:14
ISC event	389808 ↗
USGS-ANSS	ComCat ↗
Local date	October 17, 1989
Local time	5:04:15 p.m. PDT[1]
Duration	8–15 seconds[2]
Magnitude	6.9 M_w[2]
Depth	12 mi (19 km)[1]
Epicenter	37.04°N 121.88°W[1]
Type	Oblique-slip

◁ ○ ☐

The Planet speaks the language before we do. It speaks the language between the years, because it is the one spacing it out and rearranging it all. The "spacer" is the Planet's voice in physical form. Like an invisible worm. That's where you dig deepest to see the Pattern in it's natural form.

KEY YEARS TO CALCULATE:

2600 1906 2017
1809 1980 2018
1849 1985 2019
1856 1987 2020
1881 1989
1890 2000
1897 2016

2600 is a magic number!

SIGN LANGUAGE

Drugs are made by hand. Cannabis is not a drug. It is an oil. A plant oil similar to olive oil and you can cook with it. It is also a blood cleaner. Cannabis is for language, Airwaves. It creates an Echoing recording studio.

Drug in heiroglyphs:

D = Hand (Fingerprints)
R = Ra (Eye)
U = Bird (Echo)
G = Throne (Land)

Drugs are a drag.

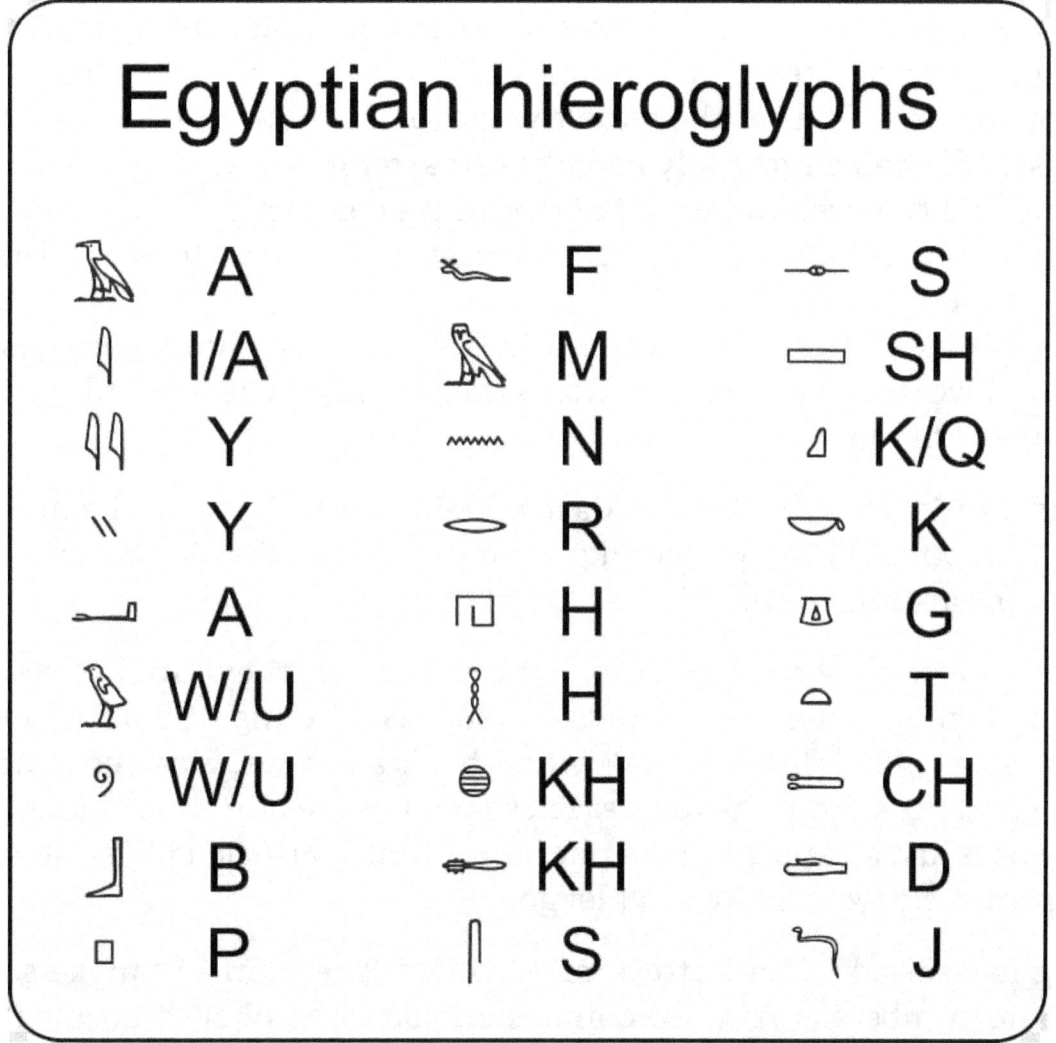

WHAT TREES ARE

Wood is the secret. Wood is the throne. Wood holds the throne. Wood projects the throne. Burn the wood and the throne shows it's face. Wood burning shows a Pattern. Ever since I began smoking every day and doing all this, the fires start every year at the same time and affect all those within my circle. That's quite the message and Pattern for me to be involved in. And they still think I'm crazy. Maybe if I wasn't in a Patterned existence that I couldn't prove and I lived somewhere less spectacular, but no, I aimed for here and here I am. So I sound weired and crazy but Jimmy crack corn and I don't care. These people need to wake up. This place is becoming very boring. I have to begin another book is what the Planet is telling me. And dissect everything knowing that Drugs are made by hand. The Egyptian heiroglyph sound of D is hand is the BEGINNING of the word. It's a vapor to read.

This existence is a bandaid and everyone is too lazy or blinded to rip it off. How could somebody ever be ungrateful? It's something I'm trying to understand. They think I'm crazy or so powerful that they don't want to deal with me. So they play games with me instead. Their games only make me smarter in the process of them building up a wall around themselves instead of opening a connection. They choose to close their blinds. I only think around and gain access to their wall to watch them as they watch me. Autopilot.
You can't outthink THINK. When you think from the Beginning, you outsmart anything. Bob and weave. Think around. That's the process to the Beginning. Keep going, keep putting the bricks on.

Hayward is the Heart of the Bay and Castro Valley is the Atrium. The upper valve of the heart. I am strategically placed like a bomb. Watch what happens next if I ever seem to disappear from view.

If people can put their phones away for a month or two, they can discover who they are. Blinding Lights, all light is blinding. Science is blinding. People are just blinded like moths to lights and flames. That's what this is all about. We need to advance the story with good news from the future. These phones are blinding, controlling magic wands. Phone is just an app. It's not a phone, it's Infinity turned physical. Attached to our fingerprints. No wonder we can't let go.

Everything is a sound in the Pattern of numbers. The numbers make sounds. The symbol of the number is memorized and then you remember to mimic that sound. You mimic what you were taught was associated with that symbol. Mirroring. Like a bird mimics what it hears. Sounds make symbols. The Planet actually speaks the sounds represented by the symbols we speak. The Planet taught us this language by sending us things to count. Birds. The Planet speaks behind the symbols of anything you see. Learn the sounds by experimenting with your sounds, then record how the Planet responds to you. Repeat specific things all day, out loud and to yourself. Write it down and then listen to the effects it has on anyone coming into your Frequency vicinity. Close or far away. The Nickelodeon will start to dance when you start command prompting. Computers were only invented to show us how this Planet works. How deep you can think is how advanced we can get. Believe you can drill electrically into the Planet and we are there.

HETEPHERES CARRIAGE

KHUFU SHIP

9-13-2020 BLOOD BROTHERS

Feeding the birds influences the Airwaves. Your then control over the waves follows you and the force field that surrounds you communicates to those within it, compelling them to tell the truth. By choosing "nicely", your force field compels their truth. Stumble upon words, magnify Echo. Especially on sacred tree ground. If you do the opposite, the negative game continues to ping back and forth and you get trapped like a hamster on a wheel. Choose to arm yourself wisely by making nice choices with your nice bubble. That's all it should have ever been. Everything is a process = ssecorp a si gnihtyrevE. Depends on how you look at something creates the Echo. And activating a super power for someone else grants you access to another of yours. This place contains extra credit hidden everywhere. It's all about briefing the others on nice choices.

I live in the Tesla Echo Bubble. It's a type of world. An alternate vision to be applied to now. Just thinking about it applies the vision. Khufu Tesla is the Master Viewer of Virtual Reality. He had control over my creation so that makes him my Maker. It all depends on how you look at your relationship with each individual warped character. Each has a new and updated story to tell. Thank you, Mr. Speak N Spell. Khufu Tesla Edgar Allan Poe. I could spot you anywhere.

This house emits bigger than Giza. It's a doorway. Anyone who enters through the front door will take on the energy of the Macabre. Force field surrounding front door. It communicates with-through Copper in the blood. It grabs you like a Megatron of force. It's heavy, strong and thick like a blanket of fog. You can't go around it, you have to go through it because it's recording and exuding through everything like Dana's walls in the Ghostbusters. Talking, telling, loudly, echoing, strongly, stronger, louder, until you put on the armour and wear it. Program it to your song. It's heavy metal after all. The Copper of the blood is in the ground. That doesn't go away. This is not a Macabre to me anymore. It was about debunking it. We got here, didn't we? We are physical now, aren't we? If you are reading this: WE DID IT! WE MADE IT TO THE FUTURE! WE DDID IT! #the experiment

Snake = Blood = J = Land = Copper = Electric

The Labyrinth would come alive for the one with the key. Einstein and Tesla were Blood Brothers. How I know this is because I was offered by the Planet a chance to do something of the same nature but with an echo of 3. Just how I was added into the equation in order to sort this all out.

Florida - 1988 - Jennifer & Unnamed friend of Jennifer (Einstein & Tesla) Echo 8

So if Einstein and Tesla arranged to meet again somewhere, they would and they did.

They even walked on a tree stump in circles together under another giant antenna tree. They are connected on the left side, so they probably agreed to poke their left hands or that's where results from your past life experiments are shown. Menkaure & Khufu. I would be Tesla's magnet and Einstein would be born through me because I contain the metallic calculation. (Copper numbers-Frequency) Encoded with the blood of Tesla and Einstein. No wonder I am so smart. My 2 Dads. I pray that one day, we will see each other as the magic metal robot copperheads we are. No more parents. Truth only. Babysitters only.

Hetepheres - Khufu - Khafre - Menkaure

Tesla and Sarah got us to Greece, but I took us all to Egypt. Free us all. The Copper in the land is the snake that will find and follow you.

The house talks and then I write it down. After Blood Brothers, Einstein became encoded with April and 4. Tesla strengthened his 8.

| 1-7-1943 = 8 + 8 (Tesla) |
| 4-18-1955 = 4 + 9 / 11 (Einstein) |

*The Planet makes the final arrangements. Copper-dirt-land-trees has wishes-echoes recorded. Think Backwards.

How Blood Brothers shows up in metallic numbers. Numbers are only sounds being said. You can see metal blood existing in the numbers. The web, the Pattern is there if you write it down. When you look at the letters or numbers, you can see them start to sparkle with the Pattern. Letters and numbers sparkle. They don't just dance, they tell a story. Just in their statuesque presence. To not see a name as a chiseled statue is to not be alive. An invisible statue that's locked with a story waiting to get out, just like the Man in the Iron Mask. Why deny yourself Infinity that you wished for? Why else be here? For me? To read these words as a reminder! Do not let them tell us magic doesn't exist if it's here. This is Storybook World, not their Monopoly game.

9-12-2020 CREATING THE STEERING WHEEL

I want to show how each scenario is a calculation. We haunt the places we go. We get to go back and exist anywhere we have ever been. This is how magnets attracts to magnets and everything is a magnet. It's a giant magnetic conveyer belt. Going around and around.

An insect flying by is not a bug but a reminder. If it is there to see: Begin. The Process. The Planet speaks dashes. Dashes are Frequency like Dahshur. Dance with the Six. The Six is Sounds. Surround Sound Recording Studio. The SOUND, not the number. Sssss-I-Kssss - SIX SURROUNDS US - Say: SIX (previously Matrix)
SIX is what we call the Echoing cup we live in. We are at the bottom of the cup and everything up and out of it is an SOS call. Understanding we live in a cup. Little Super Guy. We don't talk, we sing!
We came here to make it talk and sing.

We should be aware that what we say are sounds that Planet processes. We should speak our requests thoughtfully and closely observe the Echoes that each process invokes. Limiting speech is the way. "What will come from my speech?" is what we should be thinking. We should speak a careful/Planet smooth language. Our words are equivalent to peanut butter. Do you speak smooth or crunchy?

8 is the deciding factor. You get to choose if you want either/or by playing Scientist all your days. You'll see how your choices begin to dance around you. You'll become the Hunter when you spot negative Patterns by observing the words and rythms that enter your force field. Ground control via Horus Vision. Learn to see everything just by listening. Be the bird and listen for the specific communication. Mute everything else like the TV show that it is. When you strengthen the Focus muscle, you appear on top of the skyscraper as they are knocking below at your door. Nobody can make you do anything. Stay home and never answer your door unless it's the Special Knock. Nobody can make you do anything. Kick them out if they disagree. You'll come out with Warrior armour installed via the Planet. The Planet tracks, follows, activates you as necessary. Your shield will directly pop up and they will back down. Your actions will be via the Planet and their Planet connection recognizes that. They will back off like a good dog that listens if you respond like the Planet would. You have to become the Planet and understand they are also the Planet before you can get the Echo/mirror to talk. Eventually the sounds of familiarity will start to peek out from their shield. Truth always shines through because Planet is MOM. Mumbling, repeating.

I didn't come to make friends. I was told to expect my friends that wouldn't forget. I came here to party and dance to take over the world. I came to do SOMETHING. I've accomplished at least one task and that's remembering. Who else can do that? Who else can remember if I can? I need everybody on the Planet to wake up already. It's time. This is the Beginning of Time.

The Conversation I Hear:

Them: Are you gonna forget?
Me: No, are you gonna forget?
Them: No, are you gonna forget?
Me: I REMEMBER! Where did they go?

9-8-2020 DJEDEFRE

Dynasty 1988 Florida: Djedefre Derek = A group of us taught him how to swim in a construction lagoon with a large wooden raft Jason = Bikes and video games, 2 skating dates in 1 night story 2605 Independence Drive, Jacksonville Beach, FL, Duval County

*The Planet will find you the most exact foster home.

Hetepheres friend. Dejavu. Vision. Jesus. New name, Jason. Born in the Bay Area 1985, also connected to Florida.

9-7-2020 BUILDER OF THE SPHINX

What are you in the process of thinking of when something happens? I was time traveling in the mirror when the earthquake hit. I was deciphering Khafre's story as the power went out 3 times yesterday. Khafre showed up and I told him he was Khafre. He said that there is a spaceship buried under the house.

KHAFRE THEN

KHAFRE NOW
WAYNE KEITH WELKER
"WILLIE"

9-6-2020 MEMORY GAME - WRITING ON WOOD

You burn it into the wood when you look at it. Vampire/Activation comes through electrical charges. Static hands that zap others, light switches, toilet flushes, closing doors, everything. The louder it gets, the higher the Frequency transferring. Electric Railroad. Nikola Tesla and his "one zap" at the lab. You might think that was just one zap, but it was a story entering. He activated the Vampire Star story by his wishing and pleading. It worked. You can understand Sarah and Tesla as the Planet becoming physical in both male and female form to leave this specific story behind. It's why we are here now and not then and this story is so precise. Even Edgar said it when he penned Dream Within A Dream. Another perfectly sewn in character. All they ever did was invent inventions to make this time easier. Everything they made has created this future we live in. This is THEIR dreams come true. It's a crash landing, but we are still here to talk about it. That proves that the Planet backs up our story.

When the Nile knew the Pyramids:
-Tutankhamun existed 1800 years after Khufu/Hetepheres

-The Queen of Oysters card I found from downstairs has the Queen sitting on a small chair, holding an oyster. Hetepheres has small chairs and Alabaster.

11:44am: Earthquake, I was in the bathroom, I ran into the hallway *I've been researching and Hetepheres thinking today

11:43:57
11:44 = 10 = 1
3.4 = 7
37.720 N, 122.125 W

5:45pm, 6:20pm, 7:26pm power goes out

7:40pm: Red Tesla drives up hill

Khufu, Khafre, Menkaure = 5, 6, 8 = 10 = 1

At the time I said "He was looking for me", the electricity shut off in the house for a second or two, then turned back on.

I was referring to Khufu looking for Hetepheres and her misplaced body. He saw me misplaced, sleeping on the couch in the kitchen. The body of my old self as an experi-

ment is under this kitchen, near the front door. So is that where she is? Under the front door of the Pyramid, where no one would look?
The river that flowed through this exact spot is equivalent to the Nile being moved and sold and used for money.

No more money or money system stories.

Everything reverts to Ancient. HANG ON PEOPLE!!!

ECHO BASEBALL: "Discovered in Cooperstown, New York, Summer 1839"

White w/Red Stitching, 4 bases, 2 teams
"Hit it out of the park" Shooting Star

Birds are wealth. Land required, no money. Birds come, knowledge surrounds. Feed the birds! Every day!

Khafre is Willie. Khafre was not a nice ruler. Khufu/Khafre were half brothers but Khufu was the Mastermind and built the first Pyramid. Then Khafre had a son, Menkaure, and they both built a Pyramid. 2 1/2 brothers, 1 son/nephew Khafre also built the Sphinx.

Khufu became Napoleon and sent a cannonball to the face of it. Then became Edgar Allan Poe writing about it. Then Nikola Tesla resurrecting the Macabre. And now here we are.

Willie as Khafre was asked by a doctor if he wanted me to have eye surgery to correct my indented eye, but he said no, thinking it would make it worse. Khafre knows his Sphinx is damaged. His Frequency exudes it. I think he just wanted to feel like the Mastermind and that's the problem. Others not being happy with just working together for a miracle cause. There should be no tiers of genius if everyone is a genius. If people understand they are special just because they exist on this Planet, that's enough. Study that. It's not about what others think of you. It's about what we do here. Focus on your Empire. We are heading for a future where we all wear glasses. All they do is add an antenna, the missing component.

QUEEN HETEPHERES

Queen, 4th Dynasty 2551
Lacey Lee Welker-Paulas (Genie) 7-10-1980

31357355553259133711998

LACEY LEE WELKER
31357 355 553259

JULY 10 1980
1337

536
71198

$$\frac{5}{8} = 4$$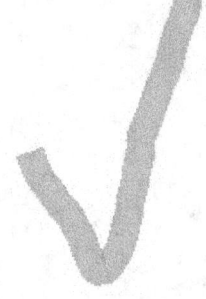

53671198

How to see all Frequency. Thinking deeply = being thoughtful.
This is how to breakdown your main Pattern magnet = YOU
Star Breakdown

Einstein found Pi within his birthdate. Where to Begin.

$$\frac{5}{8}$$

$$\underline{2600 - 2020 = 580}$$

$$\underline{1980 - 1922 = 58}$$

$$\underline{2551 - 1980 = 571}$$

$$\underline{2551 - 1836 = 715}$$

$$\underline{2551 - 1971 = 580}$$

9-5-2020

Hetepheres was 49. The Throne Lady. Sitting chair. Elizabeth had a pin inserted in her leg when she was a teenager. She sat a lot and fed the Hummingbirds. She had the mini wooden throne ready for me to aim for. I still have the stool with the blue writing in the RA symbol. I used to use it to watch Sesame Street.

How the Pattern travels:

1902: Excavation started
1920: Elizabeth Lorraine Blood Welker (Carrier)
1922: Sarah Lockwood Denise Pardee Winchester (Sallie)
1980: Lacey Lee Welker Paulas (Genie Greentrees)
2600 BCE: Hetepheres

SerehPeteh = Sarah/Peter (Ptah) = HETEPHERES|GETERNERES

H = Hut/House, T = Cup/Echo/Land, P = Build/Brick/Layer, R = RA/Vision, S = Electricity

Has specific Pattern of 4 and 2

4 E
2 H
1 P
1 R
1 T
1 S

Greentrees = G = Chair symbol (Also J sound, Goggles, Gama, Gamma, Grandma) Guh or Juh

4 E
2 R
1 G
1 N
1 S

This August/September a white house 2 doors down was sold by a Realtor named Sarah. The house sold was next to my neighbor Peter's house. The Hansen Landscape truck parked outside is the icing.

2600 - 2020 = 580

1980 - 1922 = 58
2600 - 1980 = 620 (8)
2551 - 1980 = 571

2600 and 8 have everything to do with the 4th Dynasty. 8 is the Passcode to the Pyramid Days. Sarah rhymes with Chair-a. Gamma is the same word as "Camel" in Semitic "Gimel." J and K sound the same.

Hetepheres I was a Queen of Egypt during the 4th Dynasty. 2600 - 2551 = 49 (8 & 4)

This is why we are all connected by 8. VIP, Golden Ticket

Certain people contain strong Patterns. The Planet provides the Patterned scene. You look back at all Patterns and write them down to show everyone who doesn't believe you.

1902: Excavation started
1925: 3/9, "George Risner Team" A photographer of the team discovered the shaft when George went back to the US for vacation" A patch of plaster discovered, revealing the deep 85 ft shaft. "Jumble of grave goods, including a white Alabaster sarcophagus, gold encased rods use to frame a canopy or a tent, gold, wood furniture, and more."
1927: They gathered to open the sarcophagus only to find that it was empty.

2017 is when all this started, 1. And now it's 2, 2. The T, T sound is here.
Battiscombe Gunn identified inscription indentifying Sneferu. By April, Reisner identified the owner of the tomb as Hetepheres, Mother of Khufu.

Hetepheres also had a wooden funerary bed/table. The bed measures 177cm - that's translated to 5'8 and that's my height. Also included in the tomb were 8 Alabaster ointment jars, inscribed in a chest. A box with 4 interior squares, all contained organic matter but two of the squares also contained liquid. Ensuing test revealed the liquid to be a three percent solution of Egyptian natron in water, which was used in the mummification process. (Experiment: Mixed life and death together in a Pattern to stay aware by intention)

CHAIR & WINCHESTER ECHO
GRANVILLE HAMILTON WELKER
HETEPHERES
1986

THE THRONE
1981

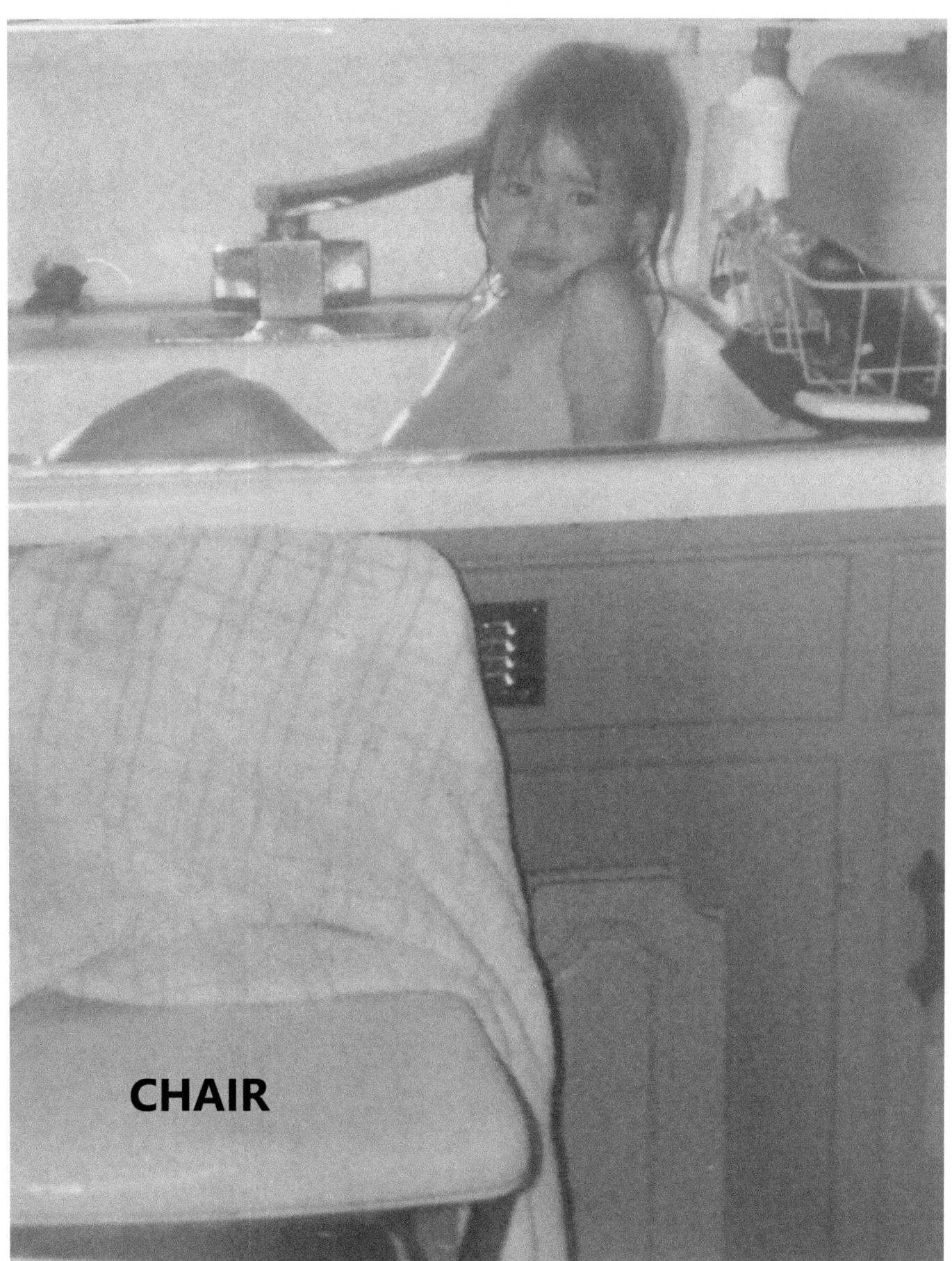

CHAIR

FOZZIE BEAR NAME TAG
1985-1986

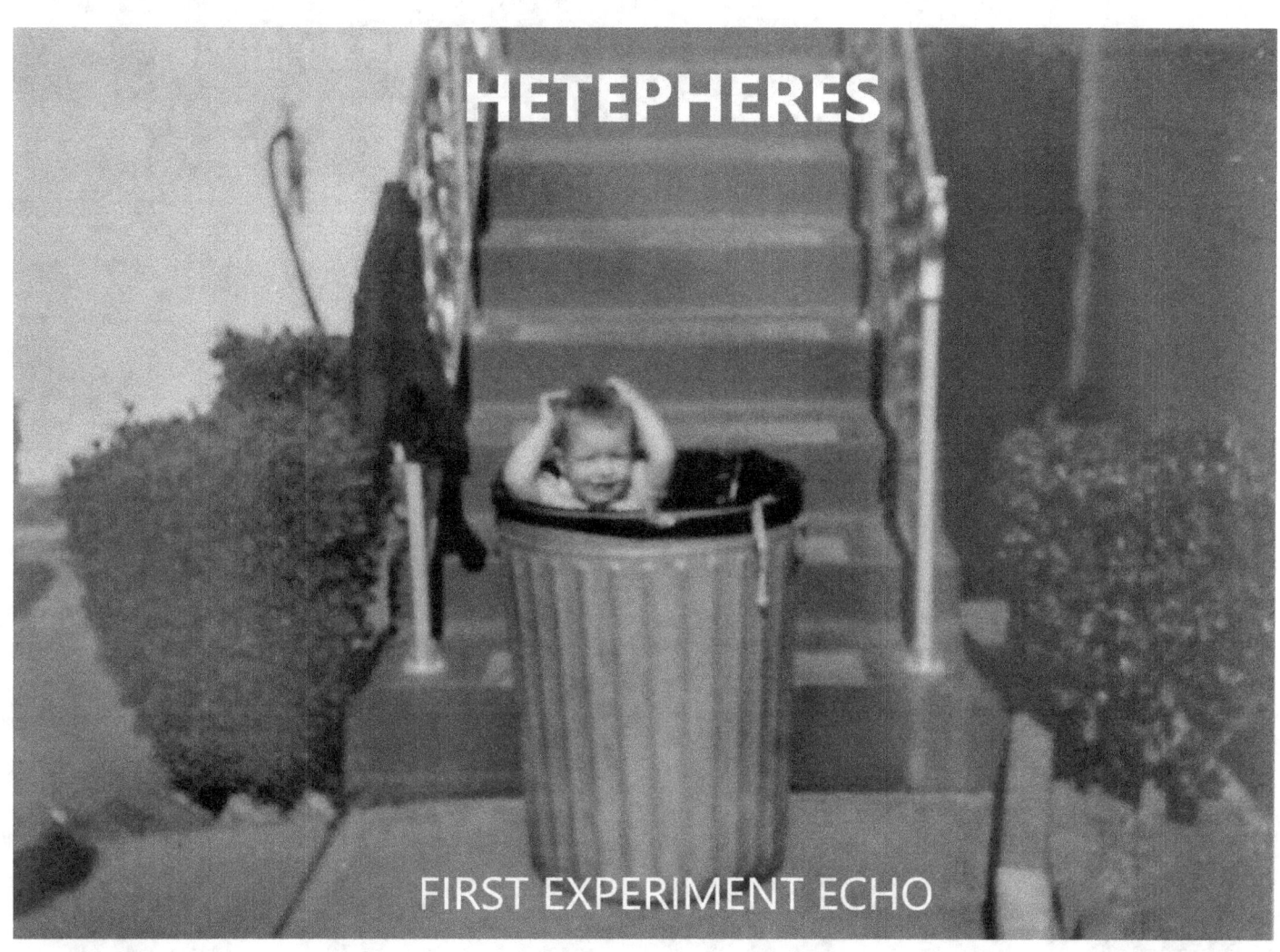

SAN JOSE, CALIFORNIA
1981

HETEPHERES
CHAIR & CARRIAGE
BABY KHUFU
1981

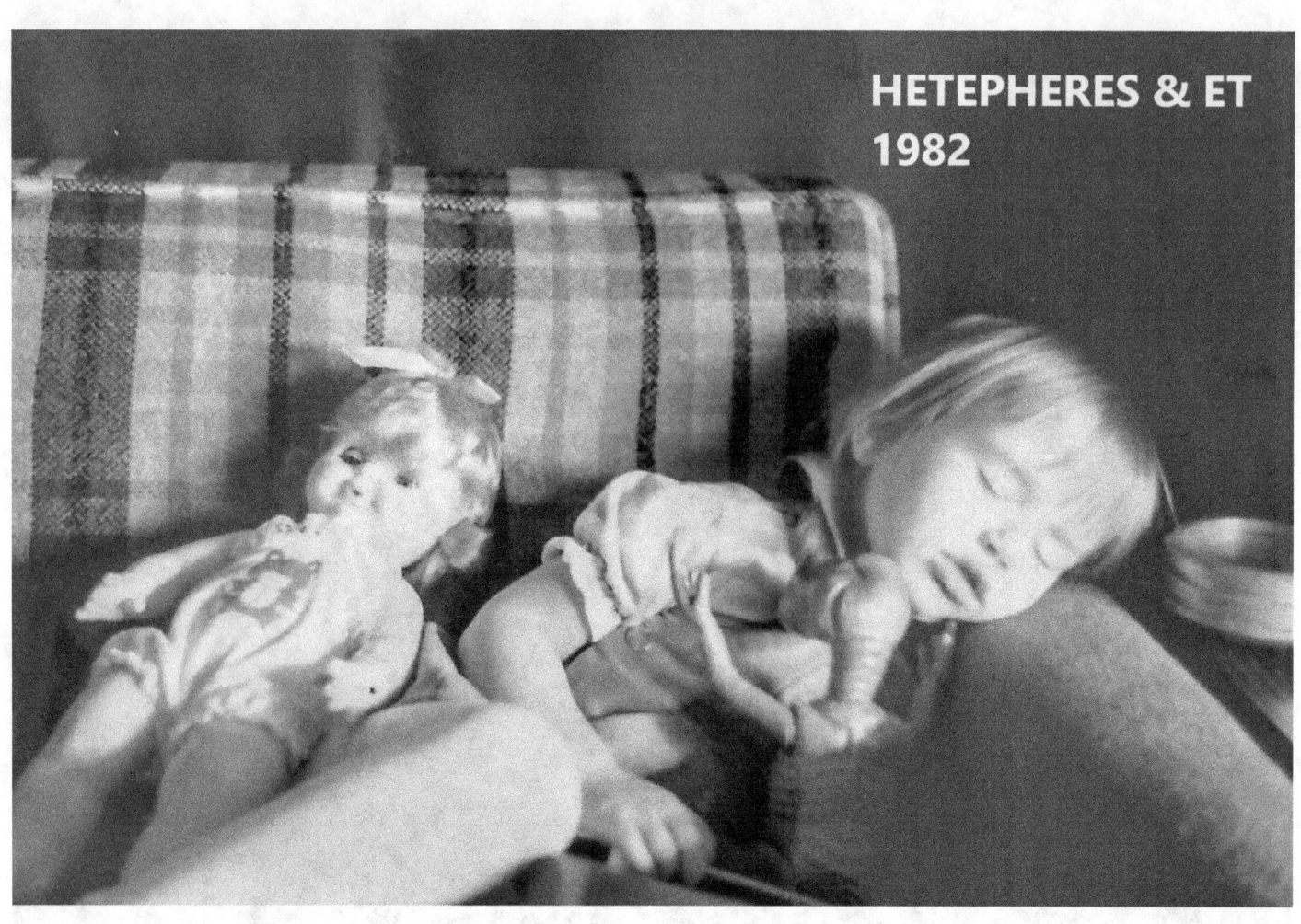

HETEPHERES & ET
1982

SAM & I
NEIGHBOR DOG
1985

FOZZIE BEAR NAME TAG
1985-1986

Certificate of Baptism

This Certifies That

LACEY LEE WELKER

Child of _Wayne K. Welker_

and _Pamella (Florine) O'Brian_

Born at _Santa Clara, California_

Date of Birth _July 10th, 1980_

Was baptized in _IMMANUEL EV. LUTHERAN CHURCH_

San Jose, California

On the _fourth_ day of _April_

In the year of our Lord _1982_

in the Name of the Father and of the Son
and of the Holy Spirit

Sponsors: _James and Pat Florine_

Rev. Kenneth E. Litw
Pastor

This Certifies

that

Pamela Jean Florine

Child of _____ Kermit F. Florine

and his wife _____ Alvern A. Florine

Born _____ September 28, 1951 _____ was

BAPTIZED

in the Name of the Father and of the Son and of the Holy Ghost

On the _____ 6th _____ day of _____ December

In the year of Our Lord _____ 1951

Sponsors: Mr. J. P. Judd

Mrs. J. P. Judd

[signature]
Pastor.

QUEEN MERITITES

Queen, 4th Dynasty
Larissa Elizabeth Esquivel (Rissa) 10-4-2001, Due 9/24

Meritites was a daughter of Sneferu and his consort of unknown name. Meritites married her half brother, Khufu. With Khufu she was the mother of the crown prince Kawab and possibly Djedefre. Both Hetepheres II and Khafra have been suggested as children of Meritites and Khufu as well and it is possible that Meritites II was a daughter of Meritites as well.

This fixes the glitches in the stories of the Beginning. They weren't dying, they were reincarnating in Patterns. All of their experiment results showed up in their future forms. The Statue of Liberty contains her blood = Copper.

"Great one of the Hetes-sceptre of Khufu" (wrt-hetes-nt-khwfw)

"Great one of the Hetes-sceptre of Sneferu" (wrt-hetes-nt-snfrw)

1. Meritites (2600) "Bloody Mary"
2. Annie Winchester (1866) "BABIE ANNIE"
3. Amelia Mary Earhart (1897)
4. Pamela Jean Florine (1951)
5. Larissa Elizabeth Esquivel (2001)

MERITITES

TWO LADIES

Patterned (Nbtj = Nebty)

Two Ladies was a religious euphemism for the Gods Wadjet and Nekhbet, two deities who were patrons of the Ancient Egyptians and worshipped by all after the unification of its 2 parts, Lower and Upper Egypt. After the unification, the image of Nekhbet joined Wadjet on the Uraeus. Thereafter, they were shown together as part of the crowns of Egypt. The Two Ladies were responsible for establishing the laws, protecting the rulers, and Egyptian country and promoting peace.

KHAMERERNEBTY & MERITITES

QUEEN KHAMERERNEBTY

Queen, 4th Dynasty
Elizabeth Lorraine Blood Welker (Betty) 10-3-1920/6-24-1998
Carmelita Isabel Esquivel (Carm) 9-21-1998, Due 9/24
Grandmother of Sarah Winchester/Genie Greentrees/Hetepheres 1980

MENKAURE & KHAMERERNEBTY

CARMELITA-KHAMERERNEBTY
ELIZABETH (BETTY) WELKER

RIGHT FOOT RETRACTED

ELIZABETH (BETTY)
KHAMERERNEBTY
RIGHT FOOT RETRACTED
STEEL PIN RIGHT LEG
"HER CHAIR" - ALWAYS SITTING
(ALSO CARMELITA)

CARMELITA-KHAMERERNEBTY
"OKT" 2000   "PO"

9-3-2020 EDWARD & THE NEWSPAPER

Macabre (In Gods Eyes)
Khufu - Sarah (Built a structure)
Nikola Tesla - Lacey Welker (Numbers)

1. Saw a black Tesla as soon as I walked into the back yard.
2. 1:56pm: Red Tesla drives down back hill
3. 2:00pm: Blue Tesla drives up back hill

Our Nickelodeon is filled with asteroids right now. Each "thing we don't like" is equivalent to an asteroid, a dinosaur, a nuissance we didn't attend to-correct.

8-30-2020 FOUR SQUARE

1. Seeing Frequency is seeing the Pattern in something.

2. When In Rome is 4 2 4

3. So when you see this Pattern, you can match it to 4-4-65 and listen to Tesla's tin can phone recording.

4. The closer you get to the Pattern, the more you adjust the station.

◆ ◆ ◆

WHEN IN ROME is 4 2 4 "The Promise"

That's how you decode things.

Michael Jackson = Billie Jean (Feet & Land) B J sounds, Planet Shoes (1-2-1983)

B equals feet and J is the Planet. Michael is walking down an illuminated sidewalk. The squares light up with each step he takes. Emphasis always on his feet. "**Moon-walker**"

Echo: Menkaure, arguing over who his son is. (Predecessor) Khafra or Bikheris Menkaure dimensions: 103.4 m base, 65.5 m height
(4-4-65 or 4-4-1906 "7")

In 1837, Richard William Howard Vyse & John Shae Perring began excavations within the Pyramid of Menkaure.

-Large stone sarcophagus made of Basalt: 8 feet long, 3 feet wide, 2.11 feet in height.

The sarcophagus was removed from the Pyramid and sent by ship to the British Museum in London but the *Merchant Ship "**Beatrice**"* carrying it was **_lost_** after leaving port at Malta on **October 13, 1838**. *Amelia was missing 100 years later*.

Menkaure (32)
Born 2532 (7, 5 = 12 = 3)
Died 2500 bce (7)

2500 - 1983 = 517
2500 - 2016 = 484
2532 - 2500 = 32
2532 - 2016 = 516

Einstein (76) Pi: 3.14159265
Born 3-14-1879 (6) (**Menkaure Dimensions: 4-4-7**)
Died 4-18-1955 (6)
Born 3-15-2016 (3-6-9) "Breinstein"

2500 - 1955 = 545
2500 - 1879 = 621
2532 - 1955 = 577
2016 - 1955 = 61

KING MENKAURE
EINSTEIN
BRIAN-BRAIN
MR. 3 6 9

Khamerernebty I

Mother of Menkaure-Einstein

Identified with the King's Mother whose partial name was found inscribed on a flint knife in the mortuary temple of Menkaure and was likely married to King Khafre. (Ryan gave me a 357 bullet)

A priest named Nimaetre is mentioned in the Galarza tomb and his tomb nearby refers to the Queen-Mother.

"She who sees Horus & Seth"
Mother of the Dual King
God's daughter
Priestess of Thoth
Priestess of Tjazepef
Great of Praises
Great One of the Hetes-Sceptre (Hetepheres, Khufu's Mother, Blood)

Maybe "Mother" shapeshifted into "Daughter" or this is the "Queen Mother Bee of the Planet"

Khufu had so many kids that when he died, he came from one of his kids. Eazy-E has a similar story.

LEFT SIDE - RIGHT SIDE
SHIP SIDES
PORT & STARBOARD
RED LIGHT/GREEN LIGHT

All focus is on left side communication. Ancient statues show this. Left side of our body is connected to our past lives. When we communicate through the left side, that's signaling a connection. For the right side, that's for this life. Reminders of this mission. Birth marks, moles, third nipples, hair, scars, broken bones. Anything that happens to either side can be linked to either this life or our past life. It's just another way to add salt to help us decode this place. It is also connected to our Tribe. Menk-aure & Khamerernebty's statue shows her left hand touching his left arm. Also, their right feet are retracted and their left are highlighted. We are vessels. Studying how a ship functions explains our canvas skin map and circuitry system.

Talking with your hands decodes the definitions of things. Antennas/wands. Our hands are telling past life or current connections. Left is a symbol for eternity/Infinity. Right is where we are now. They all have the numerical Pattern of the Pyramids embedded in their stories. They "die" - then you go chase the Frequency from their past life and match it up to now. There was a B in the name, then find the B in a newer name. The Pattern floats forward like the Bee or Butterfly. When letters and numbers in names start growing, adding on, morphing, advancing, then you can see the original beginnings of where each new letter/sound came into the mix. Billie could have been Brutus or could become Wilhelmbra, but the B is there and it's a strange Frequency to "be" present because it seems like a more advanced sound. Even the letter is strange. I guess 2 T symbols became a B. T = D, and it looks like 2 hills. That has to do with feet just as the B is the symbol for it. Beat It, B must be the Planet's story. B is very "present." High Frequency beat drum.

My Tattooed Symbols (Naming the Ship)

Left side: Lowrider symbol = Hat, Sunglasses, Mustache 1996

Right side: "Dulcius ex Asperis" 2007, Treble and Bass in a heart shape 2012, Ankh 2018

Center: "Lacey" 1994, symbol for "Mother" or "Togetherness", or up for interpretation 2008

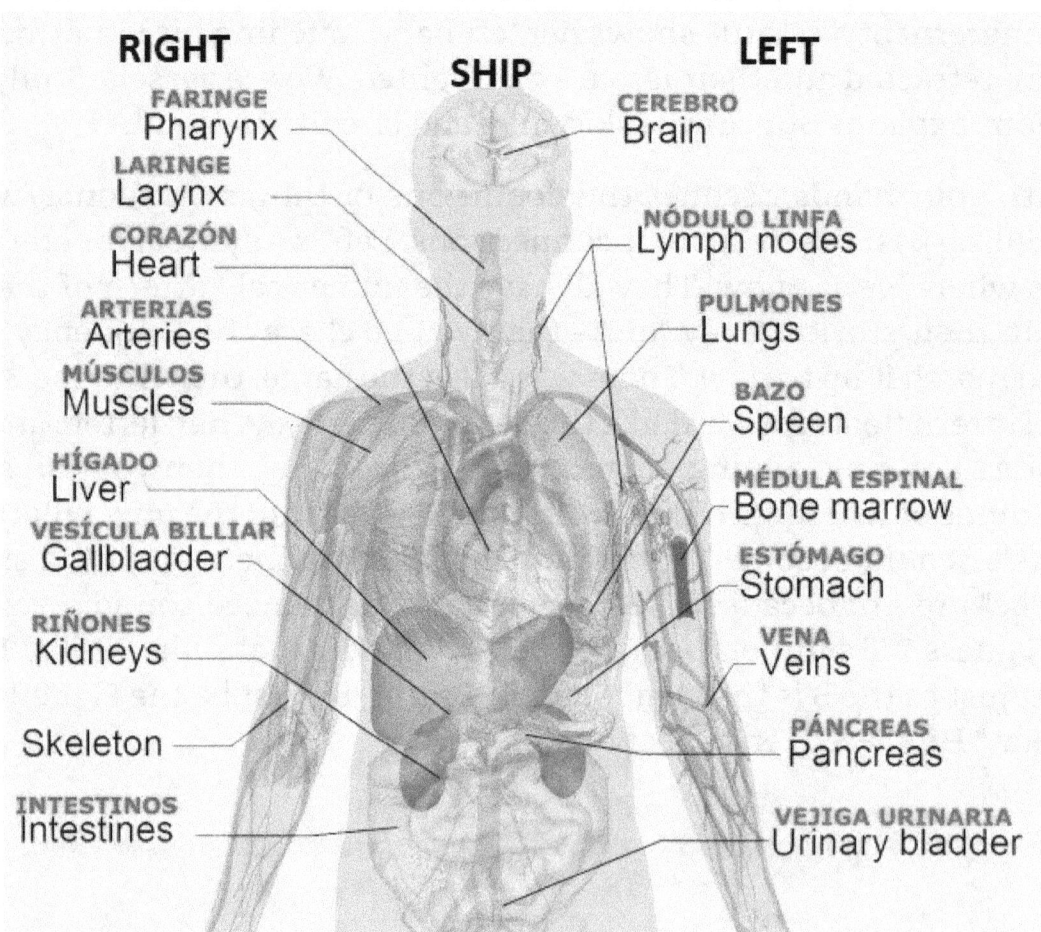

8-29-2020 SKY CAVALRY

Birds are medicine. You think of birds as just things making noise - something to look at. Calling on the birds makes everything better. It's calling in the cavalry. The true and Ancient one. The one that forms a giant serpent wave capable of genienormous proportions. An invention of the most noble kind. You would never suspect the crazy lady who talks to birds and bleeds on trees.

They came to fix things like Batteries Not Included. The original Hall Monitors. No guns required. And they get it all done by infiltrating your dreams. Want to know what made you have a vision so strong it made you change your mind? The birds are not birds. They are electrocutioners. They sit on the power and phone lines to remind you. They poop and pee where they want, owning, encoding the land with their paint DNA. Ensuring their future. Showing us how we should exist. The Planet puts the bird scenarios there for you to decode. Birds do not ever have to be in your view, but when they are, you can assure they are sent to remind you of why you are here and the invisible power you have.

Birds can avoid areas like the plague, but if they are around, you remember everything is okay. A bird is only a pillow of comfort. We rest on their wings just by them existing next to us in a tree. We count on them to be there. They remind us of why we are here and where we came from. The freedom that is ours to take back. We look at them flying and we send up a drone Frequency to try and mimic what they are doing, imagining to Infinity to see what they see. We want to fly more than anything and we are still trying to figure it out. So, you see why I see them as Master. Somehow they ended up there and we are down here, but at least we are together in more ways than one. We will figure it out.

If you see a bird, a warrior army is near and you are never alone. Imagine being stuck on an island and you see a bird after zero signs of hope. That bird electrically defibrillates you. It's not a bird. It's electricity. A jolt.

8:54am: Blue Tesla driving down hill

9:20am: The birds came by looking for some snacks. I put more sunflower seeds out. There were 3 out front. My growing orphanage.

A dove, a hummingbird, and a flock of crows. The Planet reads this. One day there will be a similar Echo on this day. Especially with the intention I'm adding in for instruction. It will be apparent as I'm planning my future as I write this out on this tree. Tree = Fortune Teller

Crows look like Corvette Stingrays, black.

When you feed the bird, you put a coin in the jukebox.

9's tell the true story of the Planet, how it works, how we exist here. The Planet uses 9s as mannequins. Anything 9 is an Echo of the original order story-instructions here. Carmelita-Catherine, Actors of the Planet. Props, too.

Thinking is dreaming. Frequency being pulled in. That's a high Frequency process. The tree encodes and records that. Where the rings come from. High Frequency = High Vibrations

Sit next to the tree every day and eat your lunch. Just you and the tree. Eventually it will get you talking and thinking. Just like the bird. I love this sparkling place. There is no thinking - it's a Process. Ancient Process. Uncover what's on your conveyer belt and learn how to use it all. Your powers are literally at your fingerprints.

Heiroglyphs: Do what the picture does: Bird = A = Do what the bird does: Say AAAA-AAAA!
It's not just an Alphabet. It's a map.

10:39am: Red Tesla driving down hill

PEEWEE & TALLYWACKER

Girl Talk happened every night for at least a little while in 1986. Peewee & Tallywacker is the story of "the talk" when I asked where babies come from. This was in Lake City, Florida, with Pamela having already morphed into Pamelia. This was also the time I learned how to "call her": "Do not call me Mommy or Mama. You can call me Mom, but not Mama!" #tumbleweed

Mama was Larissa's nickname. The Planet was teaching me to speak the correct sound to talk to IT. The nickname was a Frequency Echo story that turned up in her life as she was a "child" with an Echo of that moment. My life moment created an Echo life moment for her. The Echo gets airlifted out via all the electrical Copper, Calcium metallic sensations playing it all like the song in a dream it is. The first story wasn't of "sex-penis-vagina", it was about experimenting in communication. To create these Echoes! We need to direct our focus away from ours and others "bodies." I think we have become advanced enough to stop focusing on boobies and other parts. It isn't doing anything but delaying advancement. Cover it up or take it all off, who cares? Body parts don't matter, they are tools and that's it. If you use them for your advancement, well you always will because that's how this Echo place works. People need to get glasses in order to correct their focus.

The Pattern is structured in the first stories that you hear. The stories you hear connect the lives. If you hear this story, then you will have heard of Sarah Winchester and her mission in your past life. It's only Khufu building a Pyramid as an experimental target. A story to return to, to connect the lives, and letting everyone know that "this is what Infinity looks like." Just like Sarah in the Labyrinth when she figures out it's all "just junk." Realizing it belongs to the Planet = J, J - 4,4 = It's our story to come back to. The Planet only provides our stage and all of our equipment. If you study how to pilot the stage, you harness the controls that turn the Planet into your wild stallion. You can even access the magic tricks App. You just have to understand it's all there and it's priceless so nobody could ever buy it. It's almost inaccessible with money. The Planet requires you to give up your religion-belief in it in order to convey your truth. Money is just trees. You have to choose: Strengthen your Frequency of the trees being the Planet or trees belonging to the people. People are controlling others with the trees. Using the trees to create wood curse leashes around their necks. The money is silencing us when given to us. Secretly drowning us deeper or forcing us to walk the plank. The more we believe and receive it, the more of an addiction it becomes. It will never go away if we don't shut off the flow with a plug. As strong as relocating the Nile is what we have the power to do. Understand that force is accessible by reading a few replacement words and that power then renders any of their words neutral. You polish your shield by understanding that words are requests. Each power you give yourself and your tribe over theirs defeats their whole purpose. You move up a level in Contra.

Staying inside your tribe except to gain more knowledge is a focus that should be held. You will be seeing each other again. You should be as cooperative with your words as possible. We are all just characters trying to figure something out. Dazed and confused, everyone should be treated with child gloves. Gods of strong forces don't need to talk to anybody. Any they don't have to do anything if they don't want. No one and nothing has the power to make anyone do anything, ever. This is a Planet of Gods. Just because we have mastered all this wizardry doesn't mean it should be neutralized into "just talking" when all we are here to do is make experimental sounds that will reach into our whole future. Each day, each wake up has been about this very task! Everyone has been trying but they've been blinded. They can't see the Pattern because there are too many things to distract, so yes, light is blinding. Just like bugs can't control themselves around lights.

The Planet made Sarah and Tesla, 2 characters in this story. Carefully placed and mirrored. They were pinging off of each other like North and South poles, but across the entire country. This giant sea turtle. We now see why they call it Turtle Island. It's all a Grand Canyon....everything. The Planet is showing us that it was doing this whole thing, orchestrating it all. Their lives play out alongside each other in a perfect beat. We were still learning the controls and needed characters to play out the instructions for us to record = now.

We couldn't just trust the first story. "Science" shows we need at least 2 examples of the same story to show a Pattern. Then you can talk about it. Khufu is the Tallywacker builder. Sarah is the Peewee builder. 1 of each for confirmation. This is the Planet's version of Bill & Ted. It collected all these characters for us to give an excellent history report about. We get their true story when we follow their number Pattern. It needed to be created in order to find it-see it. Sarah and Tesla were nothing physical to us. Only stories. The Planet used their bodies to complete it's tasks and to lay the plans correctly. Numbers, not emotions. Those people were exact.

They lived in the trenches of the mind: Where the Planet does and we all do and need to in order to cure our blindness. You can only see the outlines of something if all you ride is the magic carpet dollar. To say "money and the dollar don't exist" turns the attention of the Planet your way. The birds will surround you and begin to tell your story. Their existence is our halo. You want a bird around.

11:42am: Blue Tesla driving up hill
11:45am: Red Tesla driving up hill

This world is not what you think it is. It's a bunch of sounds teeter-tottering around. Once you realize where you actually are, you can let go and breathe. Take off the belt forever and just breathe. No more anything but breath. Be. B. Bee - buzzing)))))

Each heartbeat, an oscillation sent through your B feet into the dirt we call Planet. But it's a giant brain. A giant heartbeat where every beat counts like Simon. The metal Copper encoded with each thought, sound, movement and then beamed into the electrical current. Your heartbeat is sent off like a song with each instance. The Copper triangle. Your fingerprints do the same when you touch trees, dirt, anything. It's simply complex.

Frequency Games: Chinese Checkers, Dominoes, Jacks, 4-Square, Boggle, Scrabble, Memory, Double Dutch Jump Rope, Go Fish, 21, UNO, Monopoly, Crossword, Word Search, Puzzles

This is what we do here.

Time is just a placemarker. Nothing more. A footprint tracking system. Witness. Notary.

If the Macabre exists, you have to change the name. It's a worm, a curse. A worm in the system. If you can spot the curse, you can debug it by making nice choices, replacing what's there. You can change it. The quirk that confirms this is Edgar adding Allan as his middle name. Issac Lea is "The Curse" and Edgar Poe recognized it. He changed his "calling."

From: Edgar Poe

To: Edgar Allan Poe

This is the riddle to his Macabre if we were never really here before and did all that while in Utero dreams waiting to get here, where we are now physical enough to do or feel anything about it. The Planet just passed us along and we still got to watch in light speed time. So the story is still there, we just didn't need to experience it because the Planet is a giant brain that can do anything it wants for Science. We are all donors that need to be reminded.

8-28-2020 PLEDGE THIS ALLEGIANCE

Every day is Christmas for me, although I am never allowed to play with "my" toys. I understand we belong to no one and can own nothing.

1:47pm: Red Tesla drives down hill, we were out in the Sun
2:23pm: Grey Tesla on opposite side of house
Around 12pm: Black Tesla SUV 4 door drive down hill

Crows say AAAAAA! I listened several times as they were screaming in unison - 2 of them. Hence why they created the A symbol. Probably the first one to teach us to talk. Like Paulie. Cheech Marin. Spanish white guy. We are all "Latin American." We are all Egyptian Gypsies. Every one of us are installed with Ancient sounds.

We don't own anyone. Slavery supposedly ended when we were still in the dream state with Abraham Lincoln or so the story was told. But that's not true because nobody is free and everyone is trapped like in the game of Mouse Trap. I don't know what it takes to fix all this apart from everyone publicly admitting who they are. We couldn't do this before because we couldn't figure it out without a connection to the entire Planet. We were still in the gathering process. Tesla wished for the magic screen. I wished for it when my Speak N Spell started talking to me. Infinite Search function is what we were waiting for. Thanks Google. This function leads us to our God capabilities. Accessing them in that fashion outsteps any "government agency" story because knowledge outside of the box leads us to freedom. Enables your Genius. Activates outsmarting anyone. A gun doesn't make you anything but the wrong kind of God. The Planet is the giant snake behind you. Never assume the Planet will let you take someone's life or happiness. The Planet makes you see what it wants you to see. Since everyone's existence is each a dream, you can't kill anyone. You just wake up in your own nightmare or dream the next "morning." You assume the Frequency you emit and it mirrors back at you the next day. What you put out boomerangs back the next day. That's why it's best to live in the day before. Sit safely behind the bunker and learn to move when the cannons aren't flying.

You see? Inception. A continuous Labyrinth explained. Warp zones everywhere. The walls of this place contain codes. What you program in the codes will pop up for you in surprise, later. "Later" is the "surprise." This is Surprise Planet. One person's nightmare becomes another one's reality. You never know who is "just a dream" or who is a God. Gods make strong choices. We are here to shape ourselves from jagged edged welding sparks to Gods of Lightning.

8-27-2020 SPECIFIC PATTERN

I only know very specific people. That's how the magnet operates. You either live in an specific Pattern or you don't. Your choices put you in your Pattern and keep you there if you stay on the line. It will show itself to you. Just look everywhere and write it down.

J's explained all the way:
J = Justice card Tarot = guy standing on his head

Junior = Sent me videos of him doing hand stands on chairs when he worked in San Jose at a warehouse lifting boxes and wood crates. Then he got a job working at Tesla and drove a green Corvette the last time I saw him.

J is upside down to look at the dirt.

J is the dirt is the sound of the land.
J calls the Frequency of the land to speak the story through the Speaker of the sound. Ghosts in human form: J

Speaking the sound calls the story.

The bigger your head. The more Planet communication. The Ankh is a map. A compass that doesn't move from where it was created. I don't know how far back you are able to travel, but the top part of you is communicating with the lower part of you. The two are separate on Autopilot but still working together through light force.

You do what the letters say. You do what the letters represent. The Planet makes you do what the letters represent. That is the Frequency that surrounds. The Planet tells us anything and everything. Record and Player = the Planet plays you. You don't do anything but read. The sounds of the letters attract more of the same Patterns.

When reading Egyptian symbols, you read from the direction they are facing. The original way to language shows there are at least 2 ways to read something. Adding instructions into a sentence of symbols is top secret and genius. You have to really remove yourself from this place in order to interpret anything correctly. Having a bird's eye view is crucial for survival here.

FIRST LOVE
JAIME LOPEZ DOB: **JANUARY 1980**
EDGAR ALLAN DOB: **JANUARY 1809**

POE ECHO

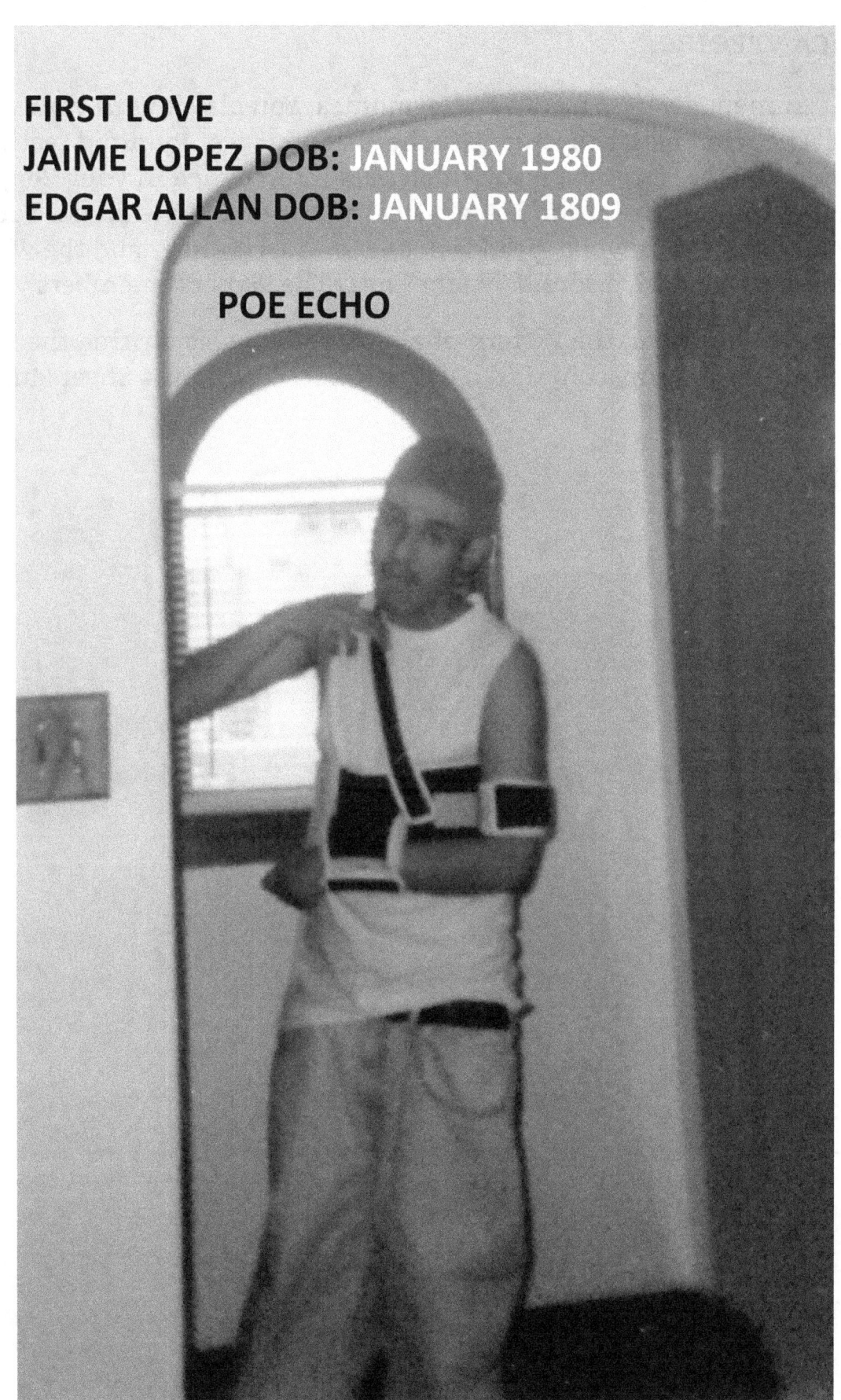

HARMONICA VERONICA

The first instrument given to me was a Harmonica. You play it with your windy cavern mouth. The wind blows through the slices of your wet Pyramid teeth. Calcium and water blowing out a rainbow song that gets intertwined in your Copper heartbeat that jackhammers the ground. A physical equivalent is the piano. You play the Pattern by listening. Your whole head is in a listening chamber and the whole focus is tuning the song to make it sound like the song. The Planet remembers you.

So writing down on paper, the Pattern of all I experience, is writing the "life" song. The instructions here. I am a physical harmonica. Instructions = Sheet Music

ANCIENT SOUNDS WE ARE

J = Land, Land Wizard, Echoing, Dirt, Data, Footprints, Blueprints

M = Mumbling, Echoes Memories, Madonna, Menkaure, Airwaves, Traveling, Moving, Mariposa

N = Free Flowing Frequency

E = High Frequency

K = King, Kween, Royal, Builder, Welder

A = Calls Birds, Loud, Phone, Voice Box, High Frequency, Talking Airwaves, First Sound, Intention

S = Sally, Snake, Water, Flowing, Serpent, Electricity

R = Ra, Sun, Force, Waves, Repeating Rays, Up (Ra-Ra-Sis-Boom-Ba)

T = Echo, Teeth, Hills, Cups, Tesla, High Frequency, Tomato

L = Lay, Sphinx, Lion, Lay of the Land, Silent Electricity

B = Baby, B to B, Foot to Foot, Step by Step (Repeats ba, ba, ba, ba)

D = Dark, Hand, Physical, Heavy, Ending

W-U = Want, Wizard, Wish, Echoing, High Frequency, Birds

F = Frank, Horned Viper, Electrical Ground Eel Serpent, Fence, Electric Fence, Force, High Frequency

*3-4 Crow flew over my head as I was figuring out "A" meaning.

WE ARE THE CIPHERS. Everything we read or see is a decoding. A memorized translation. Paper is the Ouija Board. These are just symbols being told to us, translated, we decipher.

NKOTB (New Kids On The Block): Joey McIntyre, Jordan Knight, Jonathan Knight, Danny Wood, Donnie Wahlberg. J, J, J, D, D or J, D, J D, J....they began in the 80's. I ran up the phone bill to $500 calling their pre-recorded 900 number.

Jim Henson: Follow his trail and messages.

8-26-2020 EVERYTHING IS A MIRAGE

Leech on your skin, a reminder. There is a reason I have the vision that I do. I was meant to fix things. Redirect rivers, restore harmony. The river is blood. Frank is an Echo of the mission. He sent me the River song by Eminem when I asked him to send me a song. Like Bella's asking Edward for a kiss on her birthday and he obviously doesn't want to. Then Edward left Bella in New Moon. Frank sent the song as a reminder from the Planet. I was supposed to focus on him, he was born in Egypt after all. The Planet knew I'd grab on. The most important thing is to piece the puzzle together. The more Planet troubleshooting you can complete while here, the better.

When you wake up, your future will be better. Each day, each choice you make is a fire hose to the flames. Thank you, Frank. Maybe I will see you again one day. "Back at 7" means everything to me now. It's a beautiful way to start the story. You even replaced my "night by the water" with an electrifying Frequency in the coordinates of Half Moon Bay. Like the Planet didn't hypnotize you to do all that. I came here and the Planet told me it would treat me nothing short of royal and you reminded me of that.

When Nikola Tesla plays Blinding Lights for you. What a magic trick! I'm royal on this Planet. They need to know they are, too. Royal hands, feet, bones, teeth, smile, and eyes, too! ROYALTY

ACT I / SCENE I / THE CONVERSATION

Few words: "I saw you with your green soccer ball on your leg."

Translation: Green crystal ball, Frankenstein head, just like "In Your Eyes" where she cuts off, then starts dancing with his head. The other ping is Tom Hanks with Wilson in Cast Away. That ball is now priceless. I hope he still has it. I wonder the date of that picture, although we can only imagine. It matches Khufu in every way. I didn't think another picture could ever surprise me more than 4-4-1906, but when I saw this one, my heart sank. This is what comparing pictures in the Sunday paper was preparing me for. WOW.

"You created me - whatever shall you do with me?"

Genie is green, not blue. Just like the sky is a rainbow and not blue. It's black just like the trees are when there is no light. When you acknowledge you want to play the game, the Planet plays along. It shapes your way, it shapeshifts you. It teaches you how to dance. 2-step. Square dance.

The very first place it started was someone wanting to become a Scientist and do an experiment. The wanting to do the experiment made him the Scientist. The Planet rearranged this story where someone was forced to work and then became greedy over others. Forced to work to support children he was told he had. He wanted to be free to still be a Scientist surrounded by his best friends. Why didn't they just tell that word story instead of boring ZZZ story?

We are here to learn, that's why. Forever and always upgrading. We are robotic, regenerating Babushka Dolls and we are here via dreams in Patterns. Operating through the tink, tonk vibrations of the sounds of Copper and Calcium. Heartbeat = Oscillator. Fluttering hummingbirds looking for plasma.

"Dreams" read the coordinates and then tell them to you. That's why when you travel to new places, you get pinged with dreams. Previous coordinates embedded, telling you a message. Giving you a mirage to add to your collection. As real as a memory because that is all anything is and all we are here for. Memory, memorizing, remembering. Playing the Memory Game. Marco Polo.

Echo back, don't forget: Red Rover, Red Rover, send Sally back over and the directions on how to get there. Going on a Bear Hunt to catch a Big Bear. These are all encoded within us from our early stages. The Planet did all this. This was the table meticulously set out for us Vampires. It's okay to accept the invitation to the fact that we aren't here to die and we are not going to die. The reason I did all this research was to prove this fact. Myself and countless others "should have been dead by now" but also have been "spared" by the Planet in order the convey the true story. To be confused no longer. Why hide the fact to that? The only one ever to have started pulling the wool over anyone's eyes is the one who got greedy.

The Planet isn't in the business of greed and it doesn't appreciate it either. It's in the process of cleaning house and it's only getting stronger. This is not even 2044 yet. You see? When 4 surrounds us, it's not US doing it - it's the Pattern happening and the Planet is letting us prove it's so. We are receiving the patterned communication because we are the Walkie-Talkies. 4 is the communication that the land is alive like the shadow electrified. 4 is an understanding before it is a number. You have to understand IT first, then it becomes it. In real time.

4 is a heiroglyph to anybody who doesen't understand what it means because it's not their native language. 4 is up for grabs then, meaning it is Infinite. That means it contains Infinite power including becoming the meaning of the process of the system surrounding us. 4 invokes the Beetlejuice carnival. The smack down of all Nickelodeons. "Do you smelllalalalalalala what THE ROCK is cookin'?" The Planet creates these characters and tells them what to write in the script - Okay!!! I was introduced to WWF in 1998.

The Planet becomes you - more of you at times. You become stronger then. A process of cooking you, burning you, branding you like the number cow. You become stronger - your Echo becomes louder-stronger in the 4. It's a dimension that surrounds. An oven cooking sand into glass. Fire pit. If our room stays clean and we put all the things back where they need to be, the Planet will stay a Planet and not a teenager tantrum. The Planet is a canvas. To experiment in morphing things around is to be the Scientist. To clean it up and start fresh is also recommended.

8-25-2020 THE GOLDEN TICKET

They aren't just thoughts, they are the lines to the story. If we all wrote it down, we'd figure this place out by looking at everybody's Pattern. People need to see where it's coming from. You move a river, but the Planet still has springs underneath the trail in the spot it's meant to be. It's like the Planet is having a stroke. When you restrict blood to the heart, "you die." Replace the river because it's the blood line to the Planet's heart. Steer, put a stint, but do not block. This area is the heart and Mount Diablo is the head. Like the bulls horns charging. It's now a thirsty bull. Thinking like this is a pull. How everything stands, the Planet is a giant bull hog-tied with no water but being fed an IV of what's being pumped through the springs. They need to stop controlling and hoarding all of the natural resources which are the Planet's ingredients to it's own self. So what you pick away and keep from the Planet will happen to all the inhabitants because the Planet feeds them itself and that's how we are the Planet and it is us before we are it.

We are only being granted physical access here. This is a VIP club for Stars to imagine their way to. We arranged to comeback here a long time ago and here we are. Only now, they've been building outside our original neighborhoods and what they did at the center of the Planet they have also dont at Mount Diablo. The 2 are definitely in communication. The conveyer belt goes back and forth like an invisible mirror. Because they both exist in the coordinates they do, they are both tin can phoning each other. Marco Polo.

I can't wait to be in charge like the bull. Please give me the balls of the bull.

Thank you for the dream world last night. I know Tesla was there. We were communicating the knowledge of the connection to our left side. I wish to learn about the right/left brain thinking Brett was talking about. That's what triggered that!

One day, it will be like Charlie's Angels and we all gather around in the morning for our coffee powwow and figure out how to save the world. Superheroes we are. It's so vast, I feel like Jesus in the Judas video by Lady Gaga.

When you write it down every day, you are working on the Railroad. The Planet understands the kind of tracks you are looking to build, then sends the crew your way. You can walk along the beach with your footprints and let the water take them in for next time or you can become a welder. If you write the words down every day, they will read it and never be alone or without you. What happens in the dates happens to the Viewer who understands the date. If the Planet understands what you are doing, it will amplify the communication and facets of the writing for the day will surround the reader. Read a date and find the connection. Numbers are there and meant to ping us. We are meant to connect to them. You'll never be alone or feel it ever again. You'll never look for love when you understand what you are. The Star is love. That's where we come from. Love is just a happy Frequency inputted into electricity. Heat and water mating. Sizzling through as all physical existence does. And that's all it does. That's all we do. Let's stop pretending to be some broken machine that we're not.

I knew there was something to look forward to coming here. - YODA

We are Ancient dials. The Planet has been tuning the bugs out. Perfecting our shine.

SHE BLINDED ME WITH SCIENCE

#planet

"I don't believe it! There she goes again! She's tidied up and I can't find anything!" - Planet

"Turns her eyes to me, as deep as any ocean."

"Now she's making love to me."

(When you realize the radio plays for only you.) YOU ARE GOD

The mission exists in your name. Stand by the glass and wood to receive more instruction. Inspector Gadget.

Einstein likes peeing outside. He insists most of the time. He's such a Planet Tarzan. He definitely receives direct communication and obeys. I love him. He's like a fire turtle come to save us all. He brings me life again, brings me fire. Brings us all fire, literally. The world started burning when he arrived. My Firestarter. Fire God. Fire Baby.

8-24-2020 STILL TRACKING REYNOLDS

Someone wished to be the Sun. To create it. Somebody has the Frequency as that massive black hole in the sky. Someone contains that suction.

The Sun is a crystal ball faceting in the sky. This is the dance floor.

From all my observations throughout this life, I've discovered that people like things that are funny and that remind them of childhood. My first popular comment received 298 likes and that was on Youtube. You have to really wack the mole on Youtube. Each platform attracts its own Frequency of viewers. Each Viewer has their own Frequency. Instagram is 9, so will attract 9. It's an Echo of the face of the Planet and how it works as a system. I take the guts apart.

("Reynolds" was his last word, so that would be the first place you find him. He will be on the street and in the town of a similar sound.)

FINGERPRINTS & ELECTRICITY

Eyes & fingerprints viewing and electrically communicating with your electrical existence will bring about and enhance your Nickelodeon. Your story of what you put out and what they see will become your surroundings. The clearer their understanding is of your intention, the louder and brighter the Nickelodeon becomes. Like Show & Tell. You have to show and explain your magic show. Harry Houdini. The more you share, the easier it will be to find you later. You create your target every day. Creating a target and a future echoing life of this one is what you are doing. You are programming the water life you are going to be born through again. The sonar is all ready and has always been communicating. All we have to do is keep dreaming. Doing what we are doing but with no fear. The understanding that we aren't dying, just hologramming forward like a Google Maps smear on a screen. That is what this is. The glass is sand is there to remind us. When Infinity is here, sand is a reminder. Sand is in our hands all day as the screens we swipe with our fingerprints. Also, to remind us.

The wireless phone that Tesla predicted contains a Camera app. You knew it would. The camera is the highest sought after feature. Of course it is!

This is what this place is. Magic Dream Land = Fairyland like Poe wrote.
Why choose anything more?

20 years in the future: HERE WE COME! #fullspeedahead

The Aurora Borealis poem is what Brett's fingerprints landed on when I had him let the Planet flip through an old poem book. The May Queen is what mine landed on when I asked the Planet for that days theme. Every person I talk to, it feels like I'm standing on the Yellow Brick Road talking to a clue. And I am. We all are. This is not random. Chess Planet Coordinates. We are the characters. The coordinates are the Patterned squares. But there is no size limit. Like the dog with the broom head character in Alice in Wonderland. Sweeping the floor with it's head. Do that. Be THAT dog. Doing what other dogs can't. That dog is an Echo of the animal headed God pictures on the walls. That's what we are. It's not a physical thing. It's how you turn an idea into a tool. Utilize the vision of what it's doing and become it. Morph it into shape to fit it into your Frequency and become it. Exactly like the spider and it's lunch. All you're learning is how to shapeshift into everything you learn about. Sucking out all of the knowledge until you're satisfied to move onto the next thing. An Ancient System. Drain, drain, drain, become new. No crying, just rocking and rolling.

INTERPRETING THE VAPORS

"The Aurora Borealis"

Edgar Allan Poe's poems are the Scrolls. The physical blueprints that show this same Pattern I see and that Tesla saw. The instructions are numerically placed within all his works as he received pure Planet communication from our Frequency line. It's been there, only certain people receive the communication as well as they way of life in order to do so, and to figure it out.

If you ever made a pact with the devil, you made an Echo via the Planet. And something that gongs with so much intention will be seen and come around again. Intention is BOOMERANG. Both huge words with huge Echoes. Just by the sounds the letters make. Heiroglyphs gonging, making sounds, and we think nothing will happen and it's all random. Our mouth amplifies the motor.

Be careful of the vessel you are stepping on. Remember the Titanic. This is that. Titanic was nothing but numbers. All anybody does is start mentioning the numbers. Because the Planet sent that. It was April and that's when things start cleaning themselves up. Hence spring cleaning.

The Planet is capable of creating a ghost ship story of the "horriblest" kind just so you could get the gigantic hint and symbol where people aren't "dying for nothing." The Planet will create Amelia Earhart, Big Foot, and the Lochness Monster. Create visions, mirages, dreams via its own genius communication and you still don't know where it's coming from? I pray these words melt into the electricity and reach the whole Planet.

The Planet creates these visions because it can and that's how this place works. There is no proof that the Lochness existed and there is no physical evidence that Amelia ever existed. As far as we know, she was all a dream and we've just been a part of it. It's all been about looking deeper into this invisible flying ship. Amelia was the Echo to wake up Kamal, the one who uncovered the Khufu Ship after she vanished. His thinking about flying and probably trying to solve her mystery as well, triggered the Planet to start the Khufu Ship communication process buried under the 42 singing bricks. That was 1954. It was assembled in 1985. Kamal died in October 1987, 13 years later, new Tesla was born.

The wood started communicating his wishes from back then via the rings in the wood. They still talk even if they are not growing as a tree. Singing bush.

Ancient phone calls are recorded in that ship's rings. That's not just wood, it's Patterned phone wood. The intention of that ship blares high Frequency communication at the center of the Planet, under the Sun and surrounded by reflective sand. It rests in the clouds is what the Planet sees. Dry clouds. So, we do use it, utilize it electrically. The story of the ship is embedded in the Calcium and Copper. All minerals because it's shared through the humidity river. The trees contain Calcium and we are mostly Calcium so the story is magnified in us because we have so much. It went from the sounds used to program it to echoing visions throughout our dreams and this world. When you see any boat or plane, it's an Echo of that. The water and trees carried the Echoes from then to here. Calcium is the remote control boat and the water is the wireless line.

Houses are also navigational beacon ships. Their direction in placement on the dirt determines answers and points in familiar directions like a fixated compass. Just because there is a door, doesn't mean it's a port side. Determine the way the house steers itself. Add lights. This ship is pointed towards San Jose. The nose is the deck and the deck is wood lifted into the air. So Planet recognizes this house as a "floating in the clouds" house like UP! This house is floating in the sky like a hot air balloon and nobody knows it but me. I feel like I've become the salt. I do not add the salt anymore because I have become it. What a way to exist if that's what this is.

Skin or this covering is made up of the same ingredients on the Planet for everyone. Skin is a blank canvas in the beginning of baking all of the ingredients. What turns out is the finished product. The Ship. The Vessel. And it's tagged and labeled with a map and buoys or pings called moles or beauty marks. They are the pings, navigational beacons on our skin. Destinations and treasure maps stretched out over these white sticks. Just as Van Gogh was a painter. The Planet is Van Gogh. Just like the Planet is each one of us. These vessels are steered and pinged around on this course, puppeted and pinging off of the Stars and the invisible ship navigating above us as well. The moles, marks, scars, fingerprints are what are controlling the Autopilot mode of these things. The Planet will put a ping where it belongs.

If you create a sound creation by intention and the intention gains momentum, the Planet will make sure it slices you on the left arm if that's what it takes to complete the story you wished for. If you wished for Twilight and Little Shop of Horrors, the Planet will ping and weld your vessel to the Frequency to magnetize you to those coordinates where you will find similar characters that match these stories. The Planet will magnetize it all together if that's what your heart and pulse beat for. What you daydream about is what's being pulsated out into the Airwaves via your pulsating fingerprints, footprints.

The fingerprints and process do the work of producing-providing the Nickelodeon. Each mole, each hair, each discoloration are caused by the sounds you ever made. All emit and all are specially and specifically designed to pilot the vessel in this realm. They weren't your parents, the Planet just provided you with some Floaties so that you can re-learn to navigate swimming here.

8-23-2020 REYNOLDS

9:52am is when I got up. I thought it was much earlier but I'm glad I slept in. I had nothing but dreams last night, but they were so vivid that I think I had them all while sleeping after the Sun was rising or I had them all night because I remember. It was here and everything was working and everything felt right. It was all moving along like it should. I know Breinstein was there. Others, too. Brett was there as Tesla helping me organize.

10:15am: Grey Tesla driving down hill, Jovy Chow's son
10:56am: Blue Tesla drives up hill

Each day is a new electrically charged layer being laid upon another. Each clear layer you can stand on and look down through to see the Pattern of the pathway, where to walk. You predict your footsteps by filling in the unknown with known Pattern. Search for it, pull it up into view, and place it on the pathway so you know where it is. You know where to step. If they don't know how to do that, teach them. Show them. Let them come to you and watch how to place our steps. Our footing is imperative. Where each step matters. EACH STEP MATTERS. Each step is numerical rainbow electricity. Every step is part of the dance. Every. Step. Billie Jean.

"Leave no stone unturned"
"Each step counts"
"Leave no footstep uncounted"
What if you woke up in the middle of the night as someone else?

Kermit Florine used to play his talk radio non-stop all day, all night. The receiver-radio was on all the time when I was there in 1993. They were the ones who offered to fly me out. The story told them to. I turned 13 in Haines City, Florida. 27 Pine Run

I'm just looking for my Mom now. I found Pamelia as Larissa, Tesla, Einstein, Van Gogh, Me, Willie, Etc. but I need to be the one to find her and I know I will be the only one who can because "I have the stories" "I lived their Echoes" - I CONTAIN THE STORY TO ANSWER ALL THE STORIES. Imagine your favorite story walking: That's Me. I can answer and fix any issue. Any obstacle. I can dry their eyes, fix their sight to see the Light. We don't have to be blind to exist here anymore. We can remove the masks that have been covering our faces. The bubble can pop and we can breathe here again. We can see because we've been rescued. Rescued by knowledge from someone who all they wanted to do was hang out peacefully with their friends, and where all those friends understand our one common goal: HAPPINESS

There is no money in that. There is no money. It's as real as a tree shaped like a

Nightmare on Elm Street book you put in your hands and read. It's just paper molded into a different type of paper airplane. One that's store bought. A story that isn't as magical as sharing with everyone and making everyone equal.

READ WHILE HIGH

Install One-Eyed Willie: High is the glue, the microscope bubble to the Rabbit Hole

To find and dissect anything:

Take out your calculator and write it down.

Tesla is pleased with this added feature, you better believe that!

1987 is when Kamal died: 33 years after discovering Shippy-Poo. 3,69

Kamal el Mallakh (Notice the center of his name is lowercase to stand out) el = L = Sphinx, 4 L's, Specific Patterns

Born October 1917, Died October 1987 "OKT" - 69, 3 days away from his 70th birthday.
Kamal died in Egypt near the 7 when I lived in 2 7's, San Jose & Florida.

TRAILER PARK

To live in an aluminum trailer is to pick it up in your hand as you are the Stay Puft Marshmallow Man calling your mother. That's your new voice when you move to a trailer. Your loud Echoing dreams will follow and blare at you like never before. You are now the bird who calls a can home. You'll never be alone alone.

Stay Puft Marshmallow Man is the communication of how gigantic Tesla made himself as an Echo, all the days he's ever experimented throughout all of his existences. Dan Aykroyd and Harold Ramis didn't know that when they were receiving the story. Harold Ramis didn't know the was the Echo of Tesla or probably didn't even know that experiment existed. You see? We don't know the story the Planet is telling us until after it's played out. That's how we need to start looking backwards. Look back at things. Look back at everything and apply this filter. "The Great Filter" That's how we decode everything: By the Echoes produced.

Each thing happens in a numerical Pattern because it's happened before in past lives or dreams, whatever we were, there was a Pattern playing out. We could never see it before because the stories were finishing up playing out. NOW - the stories are all happening again in opera singing unison. They are playing out around us, before our very eyes. The mission is to realize this and choose the absolute best decision that would be the voice of the Planet, not money story voices. Exist 3000 years ahead of where we are now and program THAT future here and now. Complete peace and Narnia, we will be flying fairy geniuses in no time!

Be the bird becoming a human for the first time: Then do what the bird would do.

"Nobody would believe the bird if he told others he was a bird." BE THE BIRD

You don't need to comprehend the lyrics or words that are being demonstrated around you. You just have to be there. The Calcium exudes our Frequency and picks up everything as a filter. Like a refrigerator is always running even when you aren't aware of it. Copper and Calcium pick up vibrations of voices, sounds around you. The beat is all we are here for. We just need to sway with everything. Sway with our bodies, sway with our heads, and sway with our voices.

The only thing anything needs to be is a soft dance. Your choice controls and amplifies the dance. Push on the gas or push on the brakes. The story is dancing as glass all around. Invisible glass computer receiving commands. The screen surrounds you and you are always on stage. Dancing is wise.

It will make you wake up when you realize you are in the middle of a movie and the

Truman Show was all a reminder just for you. Realizing you are in a story will make you cry, every day. Thankfully when you see none of it was wasted and you can look back and connect it all at anytime, it's quite freeing. You will feel your wings start to grow out of your shoulders.

Every movie was made to remind us. Every movie, every song - that's the Planet swaying you a reminder. Remember everything. It's the most important anything. The Planet is the only one talking to you about before. Only. One. Talking. This is not your Planet, this Planet belongs to this Planet and it sent you a friend request to exist here. So be kind and RSVP or begin again and hear these words.

The Calcium communicating with the Frequencies will communicate to the Planet to adjust the coordinates. Like a pilot programming your destination from the cockpit. The dispatcher exudes through you and will send you on your way with or without hurdles. Each day, each choice either erases or puts on your path. You write the story as you are living it. Each word you say already exists in a new Echo world scenario in the future. Each word sears it into existence via Echoland. What a wild place.

Writing it down proves that it is real. Writing it down predicts it. Not writing it down keeps you in suspense and disarray. Writing it down proves that you were here by ensuring your future with big Echoes. Choose to find yourself! Leave breadcrumbs! Each one of us needs to understand their story is just as important as any of their favorite stories about anyone they have ever heard. Marilyn Monroe, Amelia Earhart, Khufu. We are as big as them and our stories are magnified now and filled with these God Patterns and stories to connect with every single beautiful, creative, genius being on this Planet. School will one day be sitting at home in groups talking to and sharing everything about everything with every type of person or being on the Planet. The whole world will be transferring coordinates. Everyone will agree to this as our new school life of stories all over the globe. We will be so electrified with knowledge, we will be able to CTRL-ALT-DELETE the damage done in a month. Just by understanding others WANT to hear our stories. Each one has God stories to tell because we are all Gods. That's why WE EXIST.

Once we all understand this simple fact, we will all understand how to "make good choices" again. Realizing the Sun is our picture taking Mother, we will understand we were never alone. And that will erase everything. We are going to smile again. Promise. Promise. Promise is a master word, no repeating.

Thankfully there is no time and this is all but a dream. That's why we are here now. Why else be here if not for magic? They haven't solved anything, just made new inventions. I have ensured my future Infinity. I teach sound as language. The Planet's language.

Itchy skin, nose, hairs moving = Scarab Echo embedded in Calcium - Communication Electrical - Impulses, sparks, storage. Think Mummy movie. Building a box with sounds. Electrical eel serpent beings. Fish walking on dirt. Mud Guppies, Hush Puppies.

Write it down until you are searching the sky with your eyes, looking for words.

"FRANKENSTEIN" "If it's meant to be"

Looks like someone slammed their hands on some typewriter keys and this is what popped up. There was a Pattern so they kept it and that's how it was done. It sounds like how the story began here. It has all the sounds. 3 N's, RA, ANKH, STEIN (CUP), EN to INE, FRANK. Looks like an Earthquake coming. (Fred Gwynne, Born 7-10)

"Herman Munster"

Frederick Gwynne

Born: July 10, 1926

Height: 6'5

- 20 years after 1906
- Amelia last seen with Frederick Noonan 11 years later
- Lives at 1313 Mockingbird Lane

EARTHQUAKE MAN

Frank was the Frankenstein story warning me of all that was near. Even the "F"lood is in it. Elon says "The Great Filter." The Earthquake Man is here. Po(e)seidon parting dimensions again. I can't wait for the day when we are eating Tesla for breakfast and talking about him like God exists and it will never be weird to see a man or ourselves as God again. There will never be any questions again, because he's here now. Everything that they ever said has come true and is happening now. Even God in shoes. What will they think about their young looking, time traveling Lion King?

If you could write a letter to God, what would you say? What would you ask? What if you start getting answers? Would you like that? Would you accept Narnia as this world? Would you take my hand and leave it all behind and understand you already died and are here again? This is Heaven. Whoops, we didn't die. We ended up here again. Let's figure out a way to transfer so we never get lost and there are no more tears because we are able to solve everything. The Death Ray hit us, we are already in Heaven and this is the state of it. SURPRISE! Surprises come in hidden Airwaves.

I can read coordinates. If a pact from your past life shows itself to you.....

Tea is genius, strainers are genius, coffee makers, refrigerators, wireless magic screens, washing machine, glass, pipes, electricity is all genius. Razors and tweezers. Pots and cups and bowls. Stove. Hot water showers-baths. Fans. Candles. Toilets. Sinks. Lighters. So, so grateful for these inventions. Imagine Ancient Egypt has this: It does now.

Take all of these comforts away and there is no "now" "here in the future." We'd be back sunbathing on the Pyramids with our grapes. "Now" would still be "Ancient" because advancing is creating things that make our lives easier. There is no time, just making things and having fun doing it. That's real life. "Time" is just a fairly new number tracking system ready for upgrade. It's not something that actually exists naturally on the Planet - you have to find or "buy" a clock or watch to see the invention work. It's just some really slow changing numbers under plastic. Like the opposite of a stop watch or watching a snail walk. That's all it is. You can put a string through the little clock and wear it like a necklace. And they have and do. A pocket watch has a dangling chain attached to it. It's just a reminder that it's an accessory.

The real time that counts is all you write down and date with your fingerprints because that is all saved and forwarded to the cloud for Echo use. So write every day and be positive with your Frequency. You are conducting the future footsteps of your path by doing this. Write the blueprint of the Fairytale and it will follow you, become you, and your path wherever you go. Your fingerprints are the wands that make it so.

The wands that talk to the Genie. Tracking....tracking....tracking.....tracking....track. The heartbeat is the blip that's tracking. The blip like a wave or hurdle. Your fingerprints pulsate (out) your heartbeat. Blood Copper recorded. Metal Black Box. We sound like sonar blips in water. But you can't hear it. Sometimes you can see your whole body shake from your heartbeat and sometimes it can shake the bed or reach across the room traveling on a wooden floor board. That's how powerful what we are doing here is.

8-22-2020 ORIGIN OF "THE CURSE"

"Letter to Isaac Lea" (May 1829)
Dear Sir,

I should have presumed upon the politeness of Mr. R. Walsh for a personal introduction to yourself, but was prevented by his leaving town the morning after my arrival. You will be so kind as to consider this is as a literary introduction until his return from N.Y.

I send you, for your tenderest consideration, a poem-

"Some sins do bear their privilege on earth."

You will oblige me by placing this among the number.

It was my choice or chance or curse
To adopt the cause for better or worse
And with my worldly goods and wit
And soul and body worship it.

Anyone with the Isaac Lea Frequency can understand now.
Thank you, Edgar Poe.

Original poem: Hudibras by Samuel Butler (2-14-1613 - 9-25-1680) = 67

Hudibras = The Curse = 8 = 2600 BCE
The Remembrance

7-24-2020 THE MASK

If someone ever asks you if you want to see your face, say no.

Poe:

> "Who hath seduced thee to this foul revolt
> From the pure well of Beauty undefiled?

> So banished from true wisdom to prefer such squalid
> wit to honourable rhyme?

> To write? To scribble? Nonsense and no more?"

7-18-2020 PAPER PHONE EXPERIMENT

8:34am: A hawk flew into and sat on the Redwood tree and let me talk to him for a bit. Then he took off about a minute and a half later towards the front of the house.

I fed the crows this morning. Whole peanuts with shells. At least 4 or 5 showed up. At one point, one crow had ruffled feathers in the shape of "6 6" on his chest. I looked over and over. It looked exactly like it. Then another bird rumpled his feathers again, trying to keep him away from the nuts. They all stayed for a few minutes and worked on the shells here in the front of the house.

Smoking weed is like smelling your own pee. You don't like it at first because it shows you your true Frequency. If you don't like it: Troubleshoot - Your Frequency is corroded. It reverts you to Planet. The true child. You can understand it or not. It's just a really smart tree. All trees are genius though.

Sang "Vampire Dream" to Brett in the bathroom upstairs with the overhead fan going and the lawn mower going at 4:45pm. The Leche book is in here with me now. I have my yellow lighter and my clay pipe. I have been in here a few hours now. Willie is here throwing a fit because I wrote my book.

The bathroom starts to talk after a while. Especially when standing in front of the mirror and sink. It's a 2-dual reflection glass-sand-Pyramids of Giza-mirage-Dejavu and the sink basin is the Echo phone. Mirror, mirror on the wall.

Tesla is the Frequency of weed. You accept and understand what is or you have more learning to do. Harsh or the Planet has no hands? Choicing. 8. Changing your perspective by understanding, changes your Frequency. Understanding Planet.

What you think about Tesla shows your true Frequency. Ask someone about him, their answer will be their answer about every single answer they will ever have. Knowing the "Tesla" sound creates a gong, awakens the gong capability in our Calcium and Copper. We being to gong Echo. Becoming loud, Echoing vampires, our lies will be uncovered and we will be fed truth until we vomit truth to begin to speak nothing but truth. Tesla is truth. But if you believe in a man God and reincarnation-resurrection back then, why wouldn't it happen always as it should? And why would it end with just one existence of a man, back then? What kind of circle is that?

Universe means it keeps happening, there are no walls or boxes. Tesla and his story is nothing short of pure Planet - why wouldn't the true story exist here then and once again as it should? Did the circle of life story stop with the invention of one book? Wouldn't it need to keep upgrading itself if the only rule is everything changes, always?

If you don't like the idea of Tesla, you don't like yourself. Knowing of his story is equivalent to looking into a mirror. Anybody and everybody needs to be at the Frequency of Tesla because after 4-4-1906, everyone is encoded with his Echo. His-this capability can be turned on, accessed, because it was his strong intention to produce a world that can easily be solved. The experiment was strong enough to produce beings with Patterns that connect to his life's work. Numbered birthdates and names are proof the Pattern is in everybody, regardless of skin color or eye shape. None of it matters. Not even in the retina. Tesla wanted to look inside the retina and how it dialates itself. I feel my retina dialating itself often as it's trying to communicate some sort of signal. That was happening often in the bathroom today. Like Dejavu trying to play. I'm sure the Sphinx is sending something and we should be able to receive it.

Tesla was looking for the missing, invisible component in the aiding of amplifying Echoes: Humidity & trees

The tiniest matter, smaller than electrons exists as humidity. Breathing is exuding. Like a wet exhuming smoke of a mummified body. A thick fog - fume. Plume of breathing water like a frog-toad is all we are. A wet breathing - breath of air is all we exist through.

It's not "just breathing" when the exhaust is constantly letting off steam. Each breath is a notch lower on the tank. We are steam engines. Each breath does more than just "breathe some air." It's an entire water-air, dual system pump. Each breath of water is adhering, mating with, ecoding with the trees and birds around. That's where the rings in trees come from. Stories. The vibrations of the stories being told, imagined, and acted out. Each one being recorded. Each one creating a future Echo.

So I think all this intenion and trading of tree book antennas might just be doing the trick. We will be wearing paper hats in this dream instead of tin can phone hats-earphones. Maybe our paper phones will gain momentum. They seem to be working great so far. I know everyone is responding to every action.

At 5:53pm, I heard someone say "Marijuana!" from far away.

CASH REGISTER MOUTH

Get your cash register mouth out of my face is what you silently say to the money hungry zombies. Going after the flesh of the dead money tree is what they salivate for, you see?

Salivate is 8 is Menkaure is their language. Instead of POWERFUL, they salivate. Like hungry salivating dogs. You can choose that Frequency or that of the Hawk. Sit in the tree, silently, strongly, invisibly to 99.9% of the world, but sits and shares its time with you for 1 or 2 minutes. Do you know how much light speed Frequency is encoded in 1 minute? Dreams, vivid, long dreams happen in a pin prick. You think it's for 5 minutes or 1 hour when all it was was a chord-key played in a song. I love trees. Love the Ancient bird. Bird is just a word for Ancient land. It's just a vision for you to decode. Seeing a bird invites you to play the game.

The cat wakes up to the fact that it is in fact, still a bird, having the "cat ate me" nightmare and the bird absorbs the dream. Like caterpillar and butterfly. A nightmare of the similar kind is what got me here. These birds are just us having dreams. Sending-receiving dreams, and nightmares but just Dejavued vision. All meaning. All encompassing of every aspect of all your facets. Infinity. It matters, it's important. We need excavation.

Alice in Wonderland was all a dream she woke up into. Just like this place and you. You stumbled into the game and were told what to look out for. We just have to reverse the flow like in the Spaceballs movie. Suffice to say, we didn't even physically exist on this Planet until we first noticed and understood our hands and fingerprints. How they work and what they can do. Until then, you haven't fully activated the controls, so you are still in a hazy shade of winter. Dreaming, fully. As organized as you want to be but still sleeping. Dreams can be very organized to the point of OCD. So where are you? Sleeping still or do you know what your hands can do?

People need to stop thinking things are so hard when I can figure this out like I do. Stop being scared of your truth and let the Planet show you your double life in one instance. Joining the 2 eyes to 1 steady Frequency Horus Vision is what Cannabis does. Just sit.

7-17-2020 THE TESLA ECHO

6:36pm Grey Tesla
6:53pm White Tesla
6:54pm Grey Tesla

7:30pm: On the Munsters, they are showing the episode about a picture. The picture is encoded with numbers and a riddle. 2 men are listed in the picture. In the beginning of the show, they showed Herman setting up and taking a flash picture. Huge, old bulbs. Echo style.

Radio announcer comes on and says to stay inside. These 2 dangerous men are on the loose with guns. 5-6 & Winchester. Everyone is stuck at home, reading comic books on the couch. Then they talk about a homing pigeon = Stingray. The episode involves cops because Tesla Echo solves crimes-anything.

"They come with their own evidence." One of the cops says as 2 of the bank robbers appear to turn themselves in at the police station.

The cop did not believe the person calling in saying these people are going to come in and turn themselves in. "They are on the way" - "Sure they are." (Over the phone!)

(This happened to me Thanksgiving 2019. ACSO Deputy J. LINN said he saw NO PATTERN when I tried presenting to him. Written down on pages with references. I called for the Jenny Lin case; dispatch sent me the exact Pattern - yet this "Law Enforcement babysitter" saw nothing. You don't want to know how he responded when I mentioned Tesla's name.)

Alameda County Sheriff's Office
San Leandro, California
San Francisco Bay Area
Status: Compromised

7-14-2020 DOCK OF THE BAY

Money is a dream about paper: Trees

Khufu Ship is THE dream ship we are each encoded with: Trees

Did someone hack the Frequency of your ship?

What ship are you on? Did you catch the right one?

#checktheschedule

7-12-2020 CHILD IS PLANET LANGUAGE

Weird Science: The Missile Scene

A child is a spaceship coming OUT OF the Planet. Through the dirt, breaking and opening the ground all around. That's how much force and high Frequency the story existence of a "child" is. The Planet created the nuclear missile forming outside its womb. Fireworks, carnival show, and all.

What you choose to put in the cup of the performer will become YOUR ECHO FUTURE.

"Child" is giant oscillator. (STAY PUFT)

The Planet will show you he isn't a child. He's a rocketship. Star ship.

Existing here is understanding you are God.

BEWARE: WELDING SPARKS

10:15-10:30am:

Larissa-Amelia & Bretla came by!

Sam is his younger brother. Mark and Cindy are his parents. I shook his hand this time. All I did was blare about him being Tesla. He had a vampire dream in a big haunted house. A group of people: his 2 friends and Larissa, along with a group of vampires. In the dream, he said Larissa walked off and his 2 friends didn't believe him. He was bit by one of the vampires but he said it didn't hurt. After that, he realized he had fangs. Just like me and my bloody tooth dream.

Larissa and Brett met in July 2019 at the Save Mart store they work at like the undercover superheroes they are. This is Spiderman. They brought me thoughtful white and purple Orchids with green leaved plants. Larissa said she wanted to give me something that wouldn't die.

12:15pm: As I'm screaming Teslaaaaaaaaaaaa in my head, Red Tesla drives by.

My yellow lighter: Bretla has a yellow vape pen. Pineapple. I smoked it this time.

We are experiments. I wish I was joking, but I think he really gets it.

In his vampire dream, he said he was talking with his friend "Ray" about this whole vampire situation. That's HE'S a vampire, but every time he tried to talk to Ray, he would focus on another friend "Jayden" and say he had to talk to him about something else. He was getting mad that Ray wouldn't listen or believe what he was saying. Brushing him off. He got mad in his dream and woke up.

This is exactly me talking to Willie-Ra, him ignoring me, looking for J-Johnny instead.

3:14pm: White Tesla drives up the hill

UK Gemstone Tarot posts on Youtube, hours late. She is wearing a "NEW YORK" shirt with a Big Apple. Talks exactly about what's going on today. Even mentions maternal conflict. Yang card, Sun card. Death. Ace of Wands. Big Stick. Divine Timing. 4 Major Arcana. Half reading. She wanted to say "May." Mentions Scooby-Doo and Wayne's World. Farrah Fawcett Fringe. Bang tidy?

At 10:48am, I took a video of Tesla in the Sun on the deck by the giant Redwood tree. I told him to say: "I am Nikola Tesla", and he did. I recorded it and posted it on Youtube. He smiled. The video was 9 seconds long. He acknowledged the amplifying of the intention by the recording and the Sun.

I shook Tesla's hand in front of the house on the sidewalk under the wooden high voltage electric post. My hand was sweaty. His was not. I didn't care. I just asked if I could shake his hand and he complied. I looked right at the Sun as I did it. I said "Look Sun."

"Darksmoke & Fire" and "Jungle Fever" popped up on the She-Ra channel today and it's been awhile since they posted. Both videos: 20:09

Tesla Tower at Wardenclyffe/1901 - added to NRHP: July 27, 2018

It's on like Donkey Kong!

It's fun thinking we exist here. But you can free yourself when you understand you don't.

Thank you, Larissa, Brett, and Planet for the thoughtful Orchid eternal life reminder.

5:51pm: Blue Tesla drives up hill
8:17pm: Black Cat leaving my porch

7-10-2020 GRATEFUL FOR LIGHT/SUN

Woke up once and saw a pink and blue sky, but I went back to sleep. It was when the Sun was coming up.

A new Tesla movie is coming out. The trailer came out today. Ethan Hawke plays Tesla. Ethan Hawke is 49.

A slew of Teslas as we drove over the Fairmont hill. Black Tesla spotted at 6:36pm, on the way to get Johnny. Then the white driving school Tesla at 6:37pm. No more Teslas until the big hill hike by the Eucalyptus trees, a grey Tesla parked in that parking lot. I was talking to Van Gogh in the parking lot of the Sheriff's Department when a red Tesla drives by going towards the 580 freeway. Then another passes by as we are exiting the parking lot, the same way. At some point on the way there, there was another black one. Van Gogh and Johnny told me Happy Birthday. Carmelita sent me a text, I will have to check it. She texted as we were getting home. The Amazon Fresh delivery was late - so right on time as that's what we were. The "cart" for Amazon was "107" on the grocery bags. I had Amazon call me because the order was 5-7pm, but they were busy, so changed it on the website from 12am-8am (going into tomorrow.) The area code that called was "206" from Seattle, Washington. My "mom" Pamelia played the song "I Just Called to Say I Love You" for me in the 80's. She was last seen in Spokane, Washington. I had to get a call on my birthday and I don't like using the phone. After the call is when Larissa texted me.

The Amazon driver came in a silver Toyota Camry. He was very nice. I moved my car into the garage so he could deliver easier. He pulled in the driveway. He wore a blue mask and had a blue-grayish Amazon pullover shirt-vest. He spoke with a Latin accent. They are based out of Berkeley and it's a store. How cool. Thank you, Jeff Bezos and your Amazon invention for making this a special day in your own numerical way. Amazon is Wizard after all.

At 9:31pm: there were fireworks in the center of CV. I saw at least red ones, white ones, too. I don't think I even saw one on 7-4.

10:29pm: I figure out, people have been buying my book! I kept getting "credits" in my account but it was $ from my portion of the purchase price. <Insert happy expletives here>

I purchased 4 copies on 6-26-2020. I didn't connect the four until now. I wasn't thinking of it. The book is on the Best Seller's rank: #2,268,862 - Native American & Aboriginal Biographies #1175 - Foreign Language Instruction #5265

Sounds and looks legit! YAY! Thank you again, Jeff Bezos! I hope I meet you one day!

Odd Pages Item weight: 13.9 ounces
Paperback: 121 pages

ISBN-13: 979-8617534933
Product Dimensions: 8.5 x 0.31 x 11.5 inches
Publisher: Independently Published 3-1-2020
ASIN: B085DKW2WS

11:47pm: Whoever signed the ship will be recognized.

7-2-2020 TWO OF HEARTS

Amelia Earhart: Extra-Terrestrial

First child: Bryan O'Brien, Birthdate 6-21

"On the line": 157 337

2 different versions of the same Frequency.

8:43am "We are on the line 157 337. We will repeat this message. We will repeat this message. We will repeat this on 6210 kilocycles."

5:17pm: Blue Tesla drives up hill

Ryan Silcocks gave me a 357 bullet from his gun, told me to keep it. 2003

.357 Magnum

文A   ⬇ ☆ ✏

The **.357 S&W Magnum**, also commonly known as **.357 Magnum**, or **9x33mmR**, is a smokeless powder cartridge with a .357-inch (9.07 mm) bullet diameter. It was created by Elmer Keith, Phillip B. Sharpe,[3] and Douglas B. Wesson[3][4] of firearms manufacturers Smith & Wesson and Winchester.[5][6]

.357 Magnum

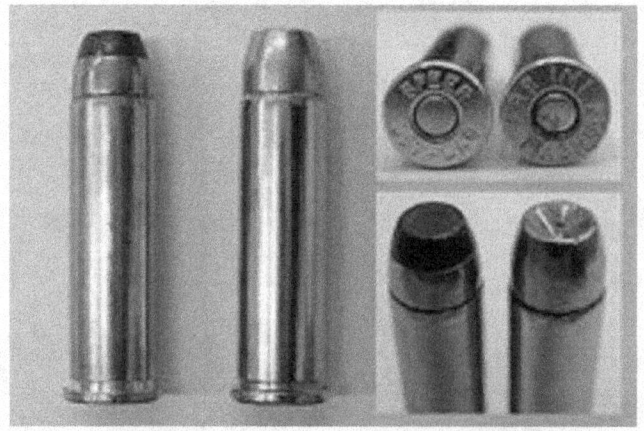

.357 Magnum ammunition

Production history

Designer	Elmer Keith, Phillip B. Sharpe
Designed	1934
Produced	1935–present

Specifications

Parent case	.38 Special
Case type	Rimmed (**R**), straight
Bullet diameter	.357 in (9.1 mm)
Neck diameter	.379 in (9.6 mm)
Base diameter	.379 in (9.6 mm)
Rim diameter	.440 in (11.2 mm)
Rim thickness	.060 in (1.5 mm)
Case length	1.29 in (33 mm)
Overall length	1.59 in (40 mm)

7-1-2020 8TH WONDER OF THE WORLD

11:24am: Red Tesla drove down hill
11:44am: Red Tesla drove back up hill

Every single thing on the Planet is silly and funny and can be decoded with an Edgar Allan Poem. Everything can be solved through his Patterning. If a true "man God" wrote a book or poems or instructions for the Planet, it would be in numerical Pattern and strong. Imagine there was no man inside of Edgar Allan Poe and Planet was puppeting him. That's how understanding he was of the "God" language he was. He was interpreting the world going on around him through his eyes. Planet blessed him to awaken to the Truth Manual. When the Planet keeps you inside and with weed, it's asking you to research. To become it.

This is the 8th Wonder of the World.
It's invisible - you can't see it.
It's language - that's invisible.
Menkaure - Airwaves
Magic Hat
Magic Language

And you have to use your imagination-understanding to use it.

6-30-2020 ELECTRICAL TREE UNITS

Where dreams come from and how we get here.

Trees are magic storage. If you think about us existing in any form, it should be in trees. Live through the tree for hours and hours, days, months, years, and you'll understand how the whole Planet works. Trees turn you back into genius. Electrical data storage, black box antenna coordinate recorders. Pretty complex.

We definitely live through-via trees.

Every day and always in movement is the system changing-clicking an entirely new stage - new wave into place. Every day is a new Frequency and a new wavelength. The whole system clicks and changes, morphs and powers, sways like ebb and flow growing a crab. It's a giant dancing metal system dangling above each one of our heads. Every day is a new episode and that's a huge deal. For some of us, each day contains magnets-scenarios big enough to write novels and movies about. That's a tiring existence. People are not the same and I am writing this to show how everyone on the Planet needs to be excused from this system "they had created" - that obviously does not work.

6-29-2020 THE GODS ARE BORN!!!

But they were dreaming the whole time and alongside the whole thing - watching it all unfold before we got here. Viewing, viewing, viewing the whole watery story until we were swept away into a daydream that we woke up from "as a child" - but no, this is not (just) a memory. This is how we enter here on this Planet. Through a memory, a first jolt memory. It had to be strong for us to remember it. What you were surrounded by amplified or de-amplified your first memory, making you remember early or later. I remembered at 2, but I was surrounded by wood and glass in my "past life" story, so of course I'd be surrounded in the same familiar back drop to wake up from that daydream into this reality. Where these walkie-talkies call "life." I see it all as a giant experiment. Each task an experiment of the grandest kind. This is what I remember being drilled into me through all my "dream communications" as a kid. To always remember and never forget. And that's why it's so loud for me. I can remember the party line dreams in my head throughout my days. Even the high and low Frequencies seeming to laugh high and low representing a man and a woman. The balance of language Frequency. Menkaure is who and what that dream was. To remember the balance. And the whole dream, each time I had it, which was systematically. None of this was physically real until we were born. Each one before 1954 is just a running story robot powered by the Planet.

5:53pm: Red Tesla going down hill

12:10pm: Blue Tesla drives by after multiple attempts at making a video about Donna and Edgar Allan Poe. I went inside to let my phone restart because it was freezing no matter what I did to try and record.

6-23-2020 23 SKIDOO

Play: Los Angeles Azules, 4-26-2018 - LIVE - Ni Contigo, Ni Sin Ti

Every day our fingerprints eminate a different color. A new song. And every day is a different color. Every year is also its own color. Everything can be broken down to a color. Color = Planet

PYRAMID ROYALE

Black guy/white Camaro - worked as supervisor at Tesla. Smoked weed with me even though he said he gets "drug tested" there. He told me I was gonna be famous the first time we went out. What a gentleman. A guy Tesla would send to treat me the way he was watching over me. He was so so kind. So soft. So loving. Thank you Tesla Planet. A true Pyramid Royal "Person." I met him on Instagram. My 1st or 2nd friend.

The other guy, Crispin Rodriguez. He was only 21. Strange soldier. Worked at Tesla after we "dated." Then came to my house the last 2 times with a green Corvette. When I met him, he only had a motorcycle. Sometimes I had to drive to the most ghetto part of Oakland to his house to go pick him up or drop him off. He liked me more than I liked him but I also know he was a nice jerk. I only attract Gods and all Gods are jerks because it's all they attract, so it's all they learned how to be. When people understand how to communicate and properly look at any any all scenes as messages, then we will start in the peace process. It's the only way.

Certain words with strong actions create new worlds. Hard, silent work towards peace is building. Being the genius in the sweat shop passing notes to the other geniuses to stand up in a wave to wash the spider out. Not one at a time. The ones standing up, getting shot one at a time, never had their army and that's why everyone "died" - to tell you to stay silent and create the silent wave of an army provided by the Planet. Work in waves and be the Planet. Silent and ninja moves. Don't open your mouth. And when you do - pay attention to the results of your experimental sounds at every interval. It's never "just talking." It's always something created to look back at to connect to; decode messages from. You understand why you meet one person from another. That's how the map works here. The Planet brings the characters and the scenes encoded with a program of numerical messages. Each scenario you experience is an experiment and should be written down to see behind the mirror.

| Egyptian | Proto Sinaitic | Phoenician | Greek |

6-22-2020

7:27am: White Tesla parked at the lake
7:36am: Red Tesla driving by Sheriff's Dept
7:50am: Red Tesla as I sat down to play outside at home

I always blow at least 4 kisses to each Tesla I see. I am a number Scuba Diver.

I saw another red Tesla driving down the hill in the back awhile ago, around 1pm.

Jovy Chow's son drove by in his grey Tesla at some point as I was trimming the slicing plant this morning.

Carmelita and Anthony dropped off weed yesterday - THANKFUL!

Einstein brought home a large tank, battery operated. Orange buttons on top. I saw another orange truck through the curtains. I see one every day now. My eye is keen for orange and it never fails to show up. My orange lighter is still working. I didn't finish the green one before I bought this one. I didn't want to run out. So using a colored lighter, fire, magnets the same color back to you. Boomerang. Or mirrors it. Fire amplifies, magnifies the color as well as the intention. The more you do it, the stronger the magnet gets. It's a muscle, you have to strengthen it all the time. It's a way to always see the Pattern. I've collected over 30 lighters I'd say. Everything is an experiment. The lady at Rite Aid taught me that when she gave me grief about wanting to choose the color of my lighter. I told her that I'm a Scientist and it has to do with everything. That shut her up. So collecting something: magnets it as a mirror to you. It becomes a part of your force field. The intention and understanding of "collecting." The Planet understands that language and will show you it's listening with responses. If you like something it has, show it what you like. If you like it's colors, show it. Maybe we've never seen color until now, this place. We are borrowing this body and everything that's physically here. These are all gifts so that we can even exist here. That's pretty much the best gift ever, to feel. To feel good, all of the time, is the aim. The target. That's why we came here, that's our mission: Happiness. And it's not hard to get it. It just takes stopping and turning around - your back to it all. Let the waves hit your back and practice your balance. Excuse yourself because that's where we truly come from. PEACE AND QUIET. Where did all this distracting noise come from?

If each house exudes 20 decibels, what if there are thousands of houses all smashed together and they call it a "city?" It's more of a deafening zombie matrix to enter at your own risk. A city and country away from noise are 2 entirely different worlds and people should be free to live and choose to live in peace. If they want to work in the city, then so be it. Cities should be used for "worker bees" - IF people want to work. The others should feel free to live the life they choose and do anything that makes them happy. That's where peace comes. Doers and thinkers should be classified as such. Understanding they are Gods, every one.

One person can create a wave of Echoes as strong as Poseidon. Some can do more, some can do less. But we all are extremely high Frequency shooting stars existing as burning, walking Suns here. The power of Echo is no joke. One person has the power to write the future down on a piece of paper or say it out loud. All it takes is understanding the Planet, communication, and what we are. Intention & Understanding are the ONLY 2 ingredients needed to create any Death Ray. 1 crowd of 100 will Echo crowds of thousands until the decibels get so loud you'll remember you're a werewolf.

6-21-2020 ALL WE ARE SUPPOSED TO DO

The Planet places you in the Pattern. Like picking you up from your shirt collar and placing you into a house, then closing the roof. That's exactly how it does it. The Planet places you in the Pattern because it already exists. It is already here and created. We are placed into it but told to focus on everything BUT what you can't see. But if you're strong and wise enough, you'll break the Patterns of what they say. All anybody is is a walking zombie robot on repeat. Sputtering the same stories to all that surround them in their Patterned worlds. We've all existed before together in the same but morphed Patterns - each time. Alongside each other in different bodies. Just looking differently but we can't remember anything. Can only remember certain things because people are "so busy." There is no time to remember when that's all we are supposed to do. That's the only reason we came. Remember. Simple. Hidden in plain sight as always. Remember to remember.

At 9:10-9:11am: with the curtains closed, I saw the blue Tesla drive by. An entirely loaded sketch just from sitting on the couch, bored, with the curtains closed. This is how amazing things happen to people. Because the Pattern IS there. If it's a concentrated Frequency-Pattern, it's going to show itself in concentrated form. Because high Frequency attracts high Frequency. Refrigerator and all it's Pyramid qualities ATTRACTS MAGNETS. Exactly like Ebony and Ivory. They are different but still the same. There is a Frequency of strength in 5 different letters in one word. So it holds the power of mastering 5 letters. Those letters are actually sounds though, so that's Frequency. EBONY - IVORY - EDGAR - LACEY

9:59am: HOLY FUCKING SHIT

The blue Tesla drives back up the hill. 49 minutes later.

I've been writing in blue.

When you write it down, the Pattern shows itself to you.

HEMATITE: Forms in the shape of crystals. It comes in various colors but also "bright rust red." It was noted that it was the original paint cover for the Brooklyn Bridge. A crystallized paint, like Ruby Slippers walking across it. So Tesla was walking and talking about this bridge. Hence why I have dreams about a bridge. Flying by a bridge near a dirty skyrise hotel in a big, old city. Even old towering brick buildings with light posts to brighten the dark night streets. I exist in the dream visions as real as a virtual reality game Nickelodeon of Tesla's eyes. His visions. Dimly lit streets glowing, towering brick structures at night. No cars on the streets. Just darkness with the light and minerals as the focus. Frequencies comparable to that of Oakland-San Francisco. I did work at Hunters Point for an armed security place at center of the projects for awhile. My shift ended around 2-3am. Then I'd drive home, across the bridge at night. I'd had dreams like that before, but it's all about making the Frequency come back to you to remind you. So you'll be put in places to make you "feel something." It's triggering you by giving you a similar Frequency of a similar place you once experienced. Electric Dejavu. It's code running code. A coincidence is a tickle of a connection. The "similar program" is running alongside, behind this one we are experiencing. The Planet is trying to mimic what you used to do. What you used to hear. The things you experienced and saw. The snapshots your eyes took and recorded. Stored for now. Explaining your unexplainable Dejavu vision. Because the word Infinity exists, there can be no limits or limiting words or laws.

INFINITY is the word of all words:

It explains everything.
It explains it all.
It explains the Universe.
It explains what we are.
It explains where we came from.
It explains where we are going.
It explains language.
It explains our mission.

The only word-sound you ever need to know. It explains ORDER.

THE BLACK CAT POEM by Edgar Allan Poe

It talks about hanging this gigantic black cat from a noose. Then a fire. It talks about hacking up his wife and burying her under the house, then the house starts talking through the bricks.

The Black Cat is talking about the Sphinx. Napoleon (Edgar Poe) already, then recently existed. The cannonball in the face. Cannon = axe. Face = head = brain. 40 years old = Edgar's life was only a dream. He was a walking, highest Frequency, number God on the Planet. The strongest Gods of the Planet will always show their presence when they are here. They will be activated and realize it. They will undersatnd it and leave a trail to follow and pick up for next time. You will find the Gods in the numbers. Even if they realize they are communicating in the Patterns or not. There will be strong Patterns visible. Napoleone Buonaparte = 10, 10 = Edgar Allan = 10

Also Napoleon Bonaparte = 8, 9 = Edgar Poe = (10-7-1849)

WHAT I SEE

Imagine stopping the Wheel of Fortune from spinning, then making it turn in the opposite direction. Then, imagine the faces of the people around you as you take control of their wheel and do something completely opposite of "what we are supposed to do" with it. Do you know how empowering that feels? I can imagine that's the power of cocaine. I ain't mad em. I wanna go home, too. Or bring home here.

Edgar loved Annie. Edgar died. The story started. Amelia was Annie. Annie was Amelia. Edgar came back as Nikola. Nikola wouldn't find Annie or Amelia this time; only hear about them and their "sad" stories. Annie would turn into Amelia this time because something happened by the hands of Edgar and now that he's Nikola, he gets no Annie. His name proves it. He knew it. He knew he was cursed as well. He stayed alone and studied alone, like Sarah. Sarah chose William. There were choices but a story about a mission of making the right choices. Choices not based off money or what everyone else is doing. Understanding you are God.

Now Nikola has found Amelia but they are now Brett and Larissa. Edgar found Annie again.
That's a circle.

On 6-20-2020, a police mannequin was found hanging from a bridge in Jacksonville, Florida. It had a pig head, like Ancient times. 94 years after Tesla's Queen Bee interview in 1926. This is a magically Patterned year.

We are not people, we are characters. CRCTRS

5-17-2020 POE DAY

Researching and missing Donna has lead me to believe we somehow die in cars. Silver seems to stand for new life coming soon. A new spaceship, craft. Aluminum supersuit.

Her interior was black. She had just upgraded from an older black Lexus to a silver one. I remember sitting with her in it. I was in the passenger seat and we were in the front parking lot of Remax In Motion at 3160 CV Blvd. But I remember noticing the new stiffness. Close windshield glass. It was stiff like a new womb/hearse. Shimmering silver water. She died in April. (4-23-2012) Her license plate: DWEAVER.

That's the answer to Edgar Allan's last poem. That is also the anniversary of losing my virginity. She had no kids, but tried. She was 43 and would have been 44 on October 7th. Her birthdate is his death date.

He was discovered in a street by Joseph J. Walker on 10-3 and died 4 days later.

"Elizabeth" was one of his poems = My Echo Grandma (DOB: 10-3)

Veronica was my nickname and his wife was Virginia.

Donna loved Twilight. Now Midnight Sun is coming out. Stephenie Meyer - McCreadie, Donna (Weaver) Midnight Sun sounds like a Poe-m of his. He talks about NIGHT, but I think he's signaling Nikaure's original Egyptian spelling. Maybe he was wondering why he had no Pyramid. He even mentioned in that poem that "no more" - "life is over for me!" O'er

He spoke of raising a tree. So he was someone connected to Khufu because the Khufu ship was discovered after Poe's death and while in incubation. I think he figured out the Pyramids in 1827: To Octavia, Spirits of the Dead - song mentions lost love Sarah Elmira Royster - The Happiest Day. (1819 - 2-11-1888)

FOR ANNIE - 1849

Nancy L. Richmond changed name to Annie after husband died in 1873
Charles B. Richmond - Rich Husband

Poe had a relationship with Nancy. Wrote poem in last year of life. Was going to publish 4-28-1849 in Flag of Our Union but changed it to Home Journal.

March 23, 1849: Poe sent poem to Nancy.

To find Annie next time, he programmed her in on that understood name to find, attract to again. Edgar died, so did Annie. His life was a dream. Maybe he was still on the ship or looking for the ship to be uncovered. He brought her back with him. Sarah's ship was built as an Echo of what was yet to be discovered in Egypt. It caught fire in 1929. When things catch fire, the coordinates are highlighted placemarkers. Sarah and Amelia triggered Kamal el-Mallakh to uncover Khufu Ship.

So Sarah was building a house while Annie and Edgar were flying around in Dreamland. Annie was Amelia. Who was Edgar? Nikola? So many options. I wish they'd all wake up.

Donna Weaver: Dreams - Decipher Book - Gift (DWEAVER)
Mitchell Nikolau: "Wake up in dreams - Important"
I met both in Real Estate.

5-16-2020 SPARKLING VAMPIRES

Hair is antenna. If you are covered in it or it's all over your face or back or fingers, it means you are royally blessed by the Planet with a tin can phone sceptre. It created your skin to be super Echous and it's equivalent to having aluminum skin or being essentially bullet proof. You can create strong, loud, high Frequency Echoes. Create waves that are stronger than the others. The antennas and Wifi pick up on you as interference. The more "tense" or high Frequency something gets, the higher my Frequency rises and things begin to break around me. The funny things is - nobody knows it but me and they wouldn't believe me if I told them, so I just laugh inside of my understanding. Talking inside of a tin can phone is having a beard or lots of hair on your egg head. What a relief.

Everything needed to break in order to see how it works and to remember the purpose of our mission. Without Napoleon and his cannon, there would be no broken faces to troubleshoot. I might have forgotten or left out some of the manual if I hadn't traced the mishaps back to the very beginning, center of the Planet. Who wouldn't think to go there? I guess I really am more Extra-Terrestrial than they are. That's annoying, but it's what we came to figure out. And a promise is a promise.

Helen Keller movie was on Light TV yesterday at 2pm.
6-27-1880 Tuscumbia, Alabama
6-1-1968 Easton, Connecticut

First job: Helene King, OCD Queen, Word Wizard, Chester Street Helen Keller with the eyes & ears.

Helen Keller represented the Pyramids Frequency communication. An Echo of Giza. Deaf & Blind. Us since we "came back" here from our creations. The toilet is just flushing the broken story it had to create to wake up. It was part of the creation here. Echo of the crash. We weren't present for any of it apart from maybe hearing it in our dream-Echo-sonar-water womb realm. Otherwise, it was just a story that never really happened to us - just the story before. While we were sleeping. We need to stop thinking anything happened to us. The truth is, we make it all happen. Remembering the mission now, erases all of what they tried to do, but that's why we came. To strengthen, to become stronger and add onto our stories. Maybe expand another Infinite world we've created but we need to fix this one in order to enjoy other existences. This Planet is the ice cream machine at McDonald's in the summer or any warm day. It needs to be fixed in order to keep us kids smiling. I think once we fix what needs to be fixed here, we will receive our communication freely once again. We will live the dream lives of Gods and we will turn the world back on. By proving we are Gods, we will show them they are, too. All it takes is understanding Ancient over money. Money is a snake charmer. The cheating drunk husband at the bar is you and the dancing masked servers are the ones feeding you more to loosen your pockets, enslaving you.

Installing a magic tree GPS locator to follow you wherever you go. Paying with gold or with teeth. High Frequency transfers done in habit - hook you on a hook, like a worm. You feel like it's sexy snake charming, but all it is, is fishing with the worms = you. They just use a little hypnotic melody to get you to jump yourself on the sharp hook.

The story of us being here is realizing the hook that has been growing through our worm necks. Waking up to the mission is the challenge. Expanding, sharing, transferring to the new future existences of us all. I feel this Planet is just a friend to communicate to. To take a vacation to. It seems as though pirates took over this cruise ship fantasy and all it really was, was a weird dream that had to happen as a map for us to look back on. This is how you create life on a Planet. This is the story it takes. An experiment, not "everything happened to me." The Planet will slap you in the face if you need it. It will give you a black eye or draw blood. Whatever it takes to show you that you aren't in charge. The programmed robot is and is powered by the Planet as well as created by. Charge: Who is in.

Once the Gods shed all of their plastic fingers and fake eyes, they will start to polish and shine for the Planet and Sun to see. The Planet wants and will set a safe enough stage for the shy Dandelions to emerge and glitter as diamonds for the world. We will and already do dazzle people, but I can't wait to have it all out loud and off my chest. I feel like I'm in labor with the Ghostbusters.

2 Earthquakes: Tonopah, NV - 4.4, 4.4, 4:27pm Reno to Tonopah = 227 miles

== PRELIMINARY EARTHQUAKE REPORT ==

Seismological Laboratory, University of Nevada, Reno

Version #726772: This report supersedes any earlier reports of this event. This event has been reviewed by a seismologist.

A light earthquake occurred at 4:27:14 PM (PDT) on Saturday, May 16, 2020. The magnitude 4.4 event occurred 67 km (42 miles) SE of Hawthorne, NV. The hypocentral depth is 5 km (3 miles).

Magnitude	4.4 - local magnitude (Ml)
Time	Saturday, May 16, 2020 at 4:27:14 PM (PDT) Saturday, May 16, 2020 at 23:27:14 (UTC)
Distance from	**Hawthorne, NV** - 67 km (42 miles) SE (129 degrees) **Tonopah, NV** - 71 km (44 miles) W (277 degrees) **Dixon Lane-Meadow Creek, CA** - 91 km (56 miles) NNE (22 degrees) **Bishop, CA** - 92 km (57 miles) NNE (20 degrees) **West Bishop, CA** - 95 km (59 miles) NNE (23 degrees)
Coordinates	38 deg. 8.5 min. N (38.142N), 118 deg. 2.0 min. W (118.033W)
Depth	4.7 km (2.9 miles)
Quality	Fair
Location Quality Parameters	Nst=23, Nph=30, Dmin=6.4 km, Rmss=0.1244 sec, Erho=3 km, Erzz=1.0 km, Gp=43.07 degrees
Event ID#	nn00726772
Additional Information	map || Did You Feel It? Moment Tensor Solutions Shakemap

For more information, see https://earthquake.usgs.gov

[Index map || big earthquake list || all earthquake list || glossary of terms || top

　Show simplified view　　　　　　　 ✕

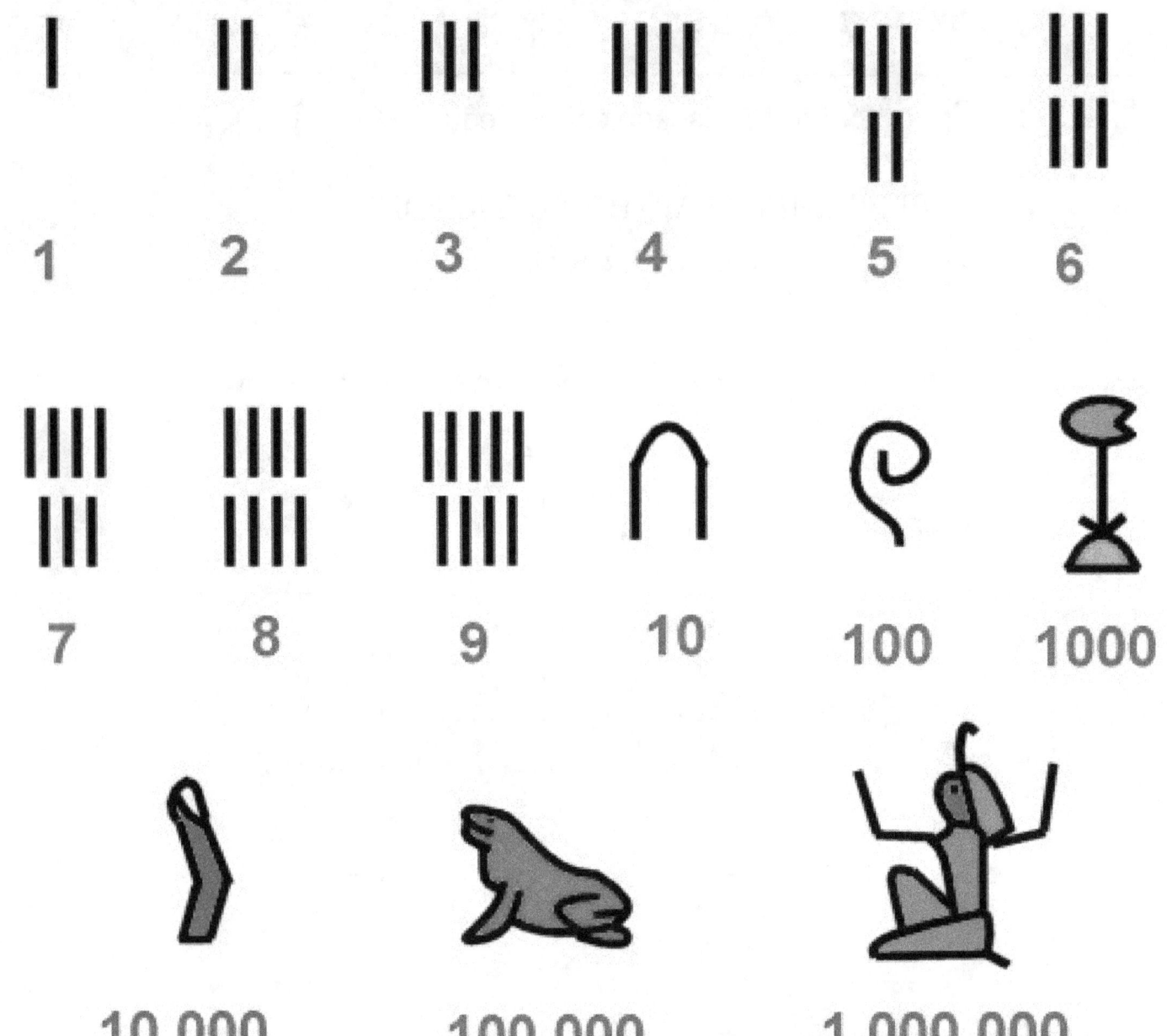

5-1-2019 MEET THE ROBOT

Superpowers in numbers. Superpowers come in codes.

We are the codes. Codes operating this robot. How do we program the same code into a robot to resurrect them? Dates, names, Patterns. Notice the strength in numbers and Patterns. Compare to the person and how they function. Observe how they respond. How does it affect them? Apply the code and you will see the Pattern. It will explain past life programming. Why things happen, why things keep repeating. It's not a curse, it's something you programmed that is showing up. Like a dinosaur. You have to replace what you programmed with something new. Making a wish in its place is what you do. That is how energy transfers and extends out to be recorded for next time. Names you once knew will show up in anagram versions. Birthdays, anniversaries, dated anything will produce Echoes for you to find in your next life. Things that are understood and recorded will show up as a new scenario in your future. With each number you comprehend creates a wave and magnet for more. It's invisible communication and infinite understanding. It's quite romantic knowing you are never alone and are satisfied with not being around others. My other half is the Planet and that's my first physical love. Everyone needs to think like this. Because it's the rule. Otherwise, we would not have seemingly impossible creations to ponder uncovering the steps to back there. Do you really think we would not have a way back? Why do people not see the Pyramids as teeth speaking language? It's that clear. So very clear and loud. It tripped me on my feet and I hit my head. I forgot the existence they told me and it's held my gaze ever since. I am hypnotized back home and it could never be diverted again. They get upset at me because they don't understand it yet. To break a train of thought that has lasted 70 years would be comparable to the death of a person's existence, their whole life seemingly wasted to them. Would their world collapse if their Frequency wasn't strong enough to receive the information? Like an old computer crashing while updating it's software. The only way they will understand is if it feeds them through airwave feeding tubes. Radio and TV stations are comparable to hospitals. You will believe anything they feed you because you think they know more than you. You forget who you are at these places because of their high Frequency. The Pyramids of Giza does the same. So does everything else. Keep your fingerprints safe holding the Ankh and you will never forget where you are again. Recorded, uploaded, shared, watched or heard on TV/radio: Intention.

We exist as kaleidoscope order. That's the key. You can see it if you write it all out, then color it. It takes thoughtfulness. That's called sitting down and actually caring enough to do it. Would you sit and write it down if it was going to save the world? Well it will. This is the key to resurrecting yourself from your past life. Things will start talking to you and you will stop feeling weird, like you don't belong or something is missing. You programmed yourself to figure this out and you left it up to someone else to figure it out - like we all did. Even me, but here I am in a house full of breaking things and I have had no choice but to figure it out. It made me write it down because that is all it was saying over and over. It was throwing things at me that really hurt, but life isn't about being fair - it's about understanding that you are the God that got you here. That is nothing but strength shoes that need to be activated in order to exist here at full potential. If a giant Moonwalker version of yourself is buried here to find and you don't stop to find it, whose problem is that really? If you heard the repeated message to build and write it down and you never stopped to fulfill the Echo, it would keep Echoing like a mechanical bull. You can compare this life to a video game, especially if you can't figure out how we exist. The Planet provides the Set. It gave us the thought for the Oregon Trail video game for a reason. It runs that game for your vision as it runs the physical conveyer belt of the numbered street you live on. It will allow and generate what appears on the street just as it gave Dan Aykroyd the idea for the Stay Puft Marshmallow Man. You're just in a new hamster wheel. Figure that out to stop the turning. It's removing the flimsy timing belt. Then you can sit there and observe, being a woke hamster. We don't have to run on the wheel if we don't want to, just because they put it there, we don't have to use it. We can devise a plan to escape from sawdust and head back to the sand. It's nice to be free.

The only thing you need to do to increase the "cognitive" ability is to add electricity. Adding Calcium, Copper, water, increasing the size of the Frequency would be Dr. Frankenstein's monster come to life. And if you did an experiment in a past life that affected your existence this time, that should be observed, recorded, and shared as Infinity and impossible will cease to exist. The definition of the word Science will soon mean 'to study' and we will all be lableled Astronaut Scientists once again. Creating a Frequency as strong as 2 people but appearing to be 1 would result in an Infinite muse. Allowing you to dream up anything you wish. You could use that as the focus to always come back to. Mona Lisa was probably smirking at the Venus De Milo. Focus. To have a muse, is to write their name on index cards and place them everywhere. You have to hypnotize the robot.

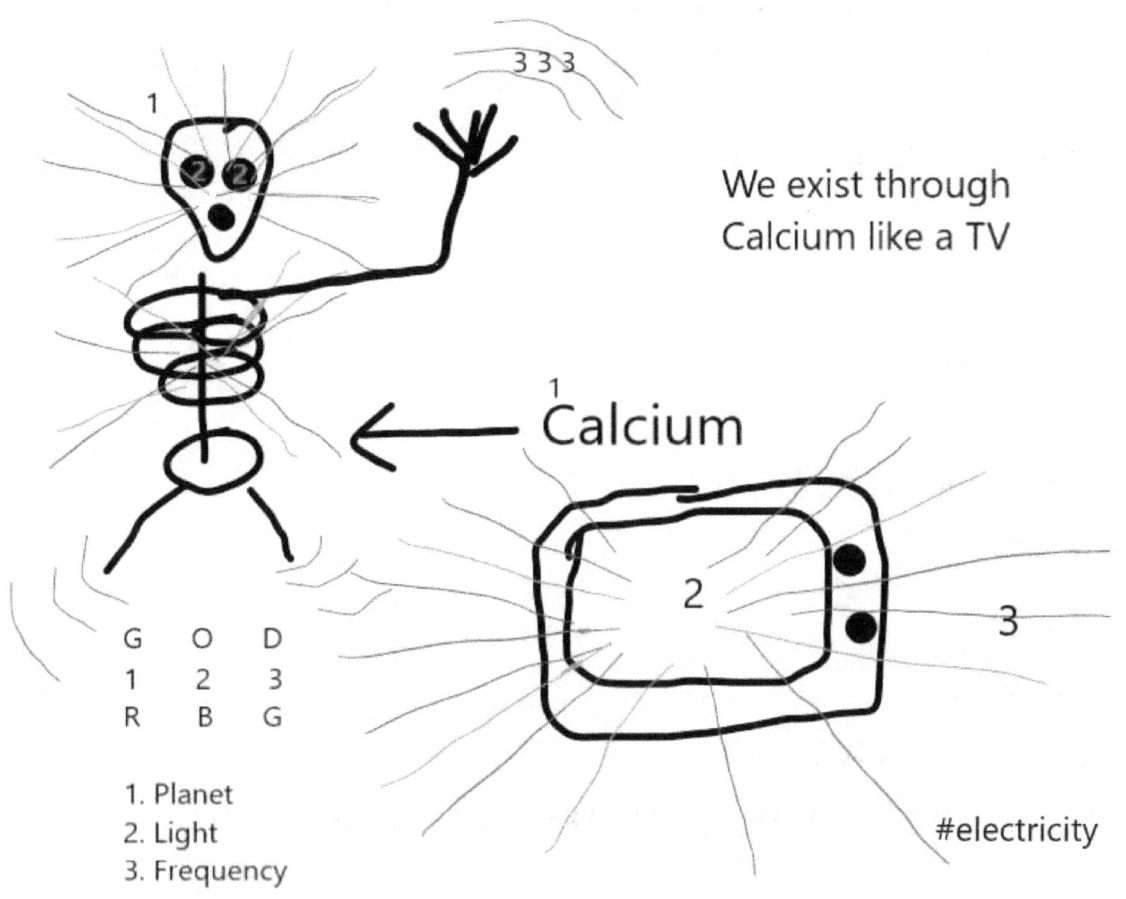

We exist through
Calcium like a TV

Calcium

G O D
1 2 3
R B G

1. Planet
2. Light
3. Frequency

#electricity

4-4-2020 MAGIC MIRROR DAY

People are guides. They will send you in the correct direction. Follow them, let them lead. Observe everything. The Planet is the one taking you on a journey. Let the people serve as directional beacons.

Nikola Tesla as Brett Bonini WAKE UP!

Khufu WAKE UP!

Larissa as Amelia WAKE UP!

Einstein WAKE UP!

Van Gogh WAKE UP!

Edgar Allan Poe WAKE UP!

Mark Twain WAKE UP!

Willie WAKE UP!

Carmelita WAKE UP!

Johnny as John Hansen WAKE UP!

NOW GONG)))))))))))))))))))))))))))))))))) AWAKEN! RESURRECTION!

Please keep cleaning the Planet and thank you for doing so. It feels really nice to be free! We are FREE LIKE BIRDS! Keep flapping! Keep swimming! We want the life we are meant to live here. Heaven on Earth-Planet. Begin Now.

3-22-2020 RING OF GIZA

Everything in my life has a weird story. I look so normal and functioning, you would not believe it if I told you. That's what happens most days. I feel like I am going around surveying the walking dead. I hope that my words stick to them like slime cannons. I'll come back later and they will have had my infection stick, right? Patience. It's about a bunch of patience. Menkaure matches, that makes sense. That's also a superpower in the form of a sound-word. It needs to be activated. Everyone is sitting at home bored, going crazy, but this is what the Planet ordered for us. This is what I have been obsessively begging for and no one will see it coming from me. 'No one is that special' they will think, because they don't understand themselves to be that special. I didn't either, but when the only words that ring from a seemingly dead mother are "You're special", you begin to look for the origination of the Echo. I couldn't stop hearing this. Now that everything's played out as it has, it even has an embedded equation for me to smile at. This is how thoughtfully you have to look back at the things that keep ringing. They aren't just thoughts. They are communication world. The invisible one that we all have to learn to break down. To see the messages sent by the Planet; these characters are made of the Planet, for you to experience and extract information from. Everyone has a numerical pattern of information that they need to share. Cast out our kaleidoscope nets to catch more fun number fish to dissect. Each time we interact with someone, it's like fishing. Once we are done, we keep looking back at the scene we created as a memory room to step into. The stronger you make the scene, the stronger the memory room will be and easier to step into. Being grateful, dropping everything, and absorbing your surroundings is all you have to do. Amplifying the interactions will lead to more snapshots being taken, creating a stronger structure.

Everything response always end back up at the same answer: We just want to be free to do what we want and this is the only way to be happy. Not having to follow someone else's rule of how to exist here. God doesn't need rules. The Planet provides everything through the help of it's characters. These are not people. They are characters with built-in SIM profiles, as are we. We are not people. Break that command prompt.

Once we realize there is a certain number program we are supposed to be running - we can remove the hijackers of our ships with the Tweezers. Remove the infestation from your foundation and rebuild the right way, with the original tool. We don't need anything but laughter and happy. This is the Frequency I've come to inject, like the bees and their honeycomb toilet maze. Stop only to squirt honey into people's honeycombs. We are here as much as the bees are. Everything that exists is as sacred and miraculous as we all are. Everything is an image of Frequency being reflected onto this screen around us. We are in the time traveling Nickelodeon already and we steer it with our minds. Just as the Khufu Ship has no ors.

Make sure to tell the passing motorbike your wishes for it is as strong as the Copper capstone when properly installed atop the Calcium bricks. Sounds are there for a reason. They aren't just noises. We have ears that hear things for reasons. The sounds ding, tink, tonk, for our invisible stone carvers to use to build our next Mecca ship body. Without the sounds, they wouldn't be building as strong. The sounds are the steel beams being personalized and welded into place. Hammers and wrenches. That's how strong and God we are. We are the tools that created tools to have all of this going on in the invisible world of each of us, yet all together. It's an endless game of Chess. We beat out and replace characters. They morph and circle in, morph and circle in. It's not death, it's hide and seek. It's Where's Waldo. Which way did they go, George? The Planet can mimic us because it created us. We live in a giant tin can phone. The K sound is the shape of a Kup - so is Khufu. The biggest message on the Planet is that this place is a cup we can speak into to make an Echo. The Pyramids are thoughtful kups to make giant Echoes. If you were a giant, you could pick one up and use it as a phone. You are and you can, just do it over and over in your mind invisible world and it works. I wish for my words to implode into the glass of everybody's home, cars, existence. I hope they drink the water that I program my words into as they pass from my drain into every faucet on the Planet. I want to be the Drano of their lives. I want to unclog this giant toilet. These hills are alive with the sound of music. Yes, they are.

3-21-2020 BILL MURRAY!!!!

I was looking for the ship.

Cory Page from Fremont took me white water rafting near Cache Creek for my birthday one year. During the ride, I tipped over at a sharp turn into the monster sized rapids. I thought I was going to die again but the last vision I caught was a group of unexpectedly placed werewolf people standing on these rocks out here in the middle of the river. I went up, then capsized into the center of the rapids. Like a giant scary endless toilet hole with water rushing all around. I didn't do this kind of thing, so I thought I was doomed. This reminds me of the Tesla rushing river story where he almost drowned but was invisibly helped off of the dam where he was stuck. The rocks seem to be programmed for us in the Frequency of 7-10 anyway. We have rock stories, rushing river stories, being invisibly helped when we thought we were doomed. I had no idea the group standing there was going to help me and that is the message. The Planet will show you high Frequency messages and it's your job to figure out what it's saying. Not figuring that out runs the doom Frequency. So, are we telling the story of what happened to the builder? Were we the builder? Are we supposed to deliver these messages to the builder? Building started in July and that's when the Sun is here and the Planet is lemon meringue. This is what's magic about this place: Just think of the number 7 to bring on the mask of the builder. 7 will surround you. Seven = sound.

Most people react to me and my thoughts laughably, unbelievably, and uncomfortably. Enough to rattle the entire interaction like an earthquake. That gets lonely, but then you remember your mission. People are too far gone, they have to figure out true happiness by letting the Planet do it for them. Forced inside by the story of a giant worm attacking the world - they choose toilet paper. Comedy is the channel the Planet is looking for. It has assumed Autopilot mode to all of our seemingly controlled remote control bodies. Your fingerprints are your remote and you have no control. It will kill only certain people and it will make it funny. Dead people and toilet paper turns into Mummies. That is pretty funny, don't you think? Can you hear the bells? Let's go home, supper is ready.

This is how I know I played a part in building them: Because they won't shut up. They sing the song that never ends. Understanding their thoughtful existence requires weed. To truly hear the song that they sang and heard, you need to put on the headphones. It's burned like a CD into the weed. So when you consume it, you begin to fly like the story of where Peter Pan came from. It's all from the same place. If it's here, you can look back to the center of the Planet and find it's origination there. You can't find it in the Greek or the Roman or even Phoenician. The beginning is the center of the Planet. All of the answers revert to the number system built there in the form of 7 structures. All you count is 7. All you have to know is 7. To get sucked in like a vacuum, play the number like a flute. It's what the Rubix Cube trained us for, to get us back to the color coded shapes. That was the Rainbow Brick Cube. Thankfully Giza is more exciting to play. The sheet music is always there, embedded. Understanding all of the ways it exists for us to decode is quite thoughtful of someone to build for us all. It's a giant thinking machine and the Thinking Man statue is an Echo because of it. It's the reason I inherited a beautiful white feathered porcelain music box with the head of Cher that plays "I've Been Working on the Railroad." I was supposed to figure out this Mad Max Madness. It's been kept in a carved wooden glass case. It has everything to do with what's at the center. Everyone needs to wake up to this fact. Be thoughtful, they are thoughtfully alive, blaring through you as much as you are them. They are our bodies sitting cross-legged programmed with all of the builder's intention. We have 4 sides, they have 4 sides. That obsessive honesty is what our bones are and where this all came from. These bodies and this Calcium we are trying to keep happy isn't ours, they are the Planet. It is it's own living thing. It's our Mother and it gave birth to each of our existences. Our parents were just the proxies that it also created. It's children being born in a giant systematic circle. It gives the final 'Go', so the final numbers are the final words of the character's poem. They are also the continuance passcode to connect to in your next game. This place is not sad, this place is not scary, this is a happy place with no words - only sounds. We are endless Poe-ts.

I looked up the Khufu Ship on Google Earth this morning. It showed the view from inside the trench, looking out. The first thought I received was Labyrinth. Brick lined walls, both sides. When you look up, a bird is flying over Khufu. You can also see Khafre. Whoever took the picture is kneeling down in the trench, towards the ground. As ominous as the bird is flying overhead, so is the kneeling towards the ground. I found Ancient Egyptian art that showed the Sun above blaring through the ship down below and buried under the ship is a white covered body. The ship is below the Sun, so where is the body? Is it in the dirt or hidden inside the Limestone? The ship was hidden by 42 bricks. That's the same number they used in the Greek Osiris story. Saying he was cut up into 42 pieces then buried in the land. But Egypt is before Greek, so the 42 was the bricks and Osiris was Khufu, not Cheops. Khafre could be Cheops. It's Kh/Ch and 6 letters. They were attempting to translate the story from the symbols as well. So where is Khufu if not buried in the ground? "I'll be at the 42 Brick Row" is what they were supposed to receive but it came out as "Follow the Yellow Brick Road."

"Don't Touch My Truck" plays by the hands of baby Einstein. So I am thinking that he wanted to be buried close to his ship, which would be like a truck. Everything built was of his passionate fire. Why wouldn't he be buried beneath it? Nobody would think to look there because it's just open dirt. Or is it? As grand as it is going up, what is below? The dirt and what you bury in it is as important as what you do above it. In fact, what you do above it amplifies what it does while it's buried. Frequency is amplified, therefore existing alive, up here, as a shadow magnet. It's the kind of thing Scientists of today are only able to dream about.

The wood ship that is built and being displayed outside of the Pyramids is helping steer everything here. It's what pulled Napoleon in, why he poetically died with Copper and arsenic in his story. Arsenic is a good preservative. It's why they chose to change his course of burial even though it was recorded on a paper will. Menkaure = Napoleon. He was being called there and became a huge part of the story as he was attracted to do so. Did they change the course of Khufu's burial and he got lost because the directions were different from what he wrote down? Did he write it down or agree to it? Napoleon appears to be Khufu lost and that is why he wrote down his burial plans and they buried him someplace completely different. That was the message of his existence as Echo of Khufu. Did he deface it because he built it? Who else would do something so loud and large? The cannon was used when it should have been "tin can" phone. It's all been about extreme communication.

Trees speak Khufu Ship. Wood, trees, ship = all one, all connected. So when you look at wood, you are looking at and communicating with the ship. It will begin to surround you like Flight of the Navigator. Vines will slowly begin to grow on your legs. You have to notice your surroundings to see the ship. Did you walk on it? Did you like it? Do you like climbing trees or standing on tree trunks? Now think of every single wood invention there ever was. Even a flute.

All the Ancients and Natives were what we were then and what we still are. What's calling us now is our communication mission here: Now that it's all set up, even the giant phone, we can make the call. It's a free call and it can change your life. The communication is coming from paper. Paper is trees. Paper will not stop talking and we will not stop talking to it. Books, rectangle paper money, toilet paper. Paper money in the form of rectangles is an Echo of the origami paper ship underneath the brick rectangle formation.

Toilet Frequency is where we are at. That is why they are highlighting it. It's how they are receiving the communication. It shows you how it's clogged. Too much toilet paper = clogs. So the clogs are speaking and telling you clearly and humorously what the problem is: Dirty, clogged, bloody, poopy, giant, public toilet Labyrinth maze Frequency. This is the signal from the channel we are receiving. It can and will be corrected by cleaning the Planet and restoring knowledge by spreading it like Baking Soda. It will unclog them when they begin to see the omen in everything. Begin to connect their dreams with here and now. The more they stay home, the more they will see the magic that turns up. Let the magic knock on all their doors. Let them all be presented with free weed. Let them research and time travel, then burst the clock by remembering we used to tell our days by the sands of the hourglass. Let's go back then. It's still here if we can throw the clocks out of the window. We live somewhere else.

My Uncle collected and worked on clocks. He never "spent time" with me. He would spend or give money. But when I sent him $50, he told me to never send him money. His character to me was supposed to remind me of all this. I was supposed to get irritated with clocks and stop believing in them. I even threw a clock of his away. It's how I froze time. Any set of numbers in my house tells whatever time the power came back on after a blackout. I see them as phone messages from the Planet. When it's 11:11 on the oven, it's not to everyone else on the Planet. It's personalized writing on the wall, just for me. I don't see it as something I need to line up. I see it as a thoughtful 'new phone, who dis?'

So I talk to the Planet and it talks to me via numbers. Anyone who would see that as insane is nothing but jealous that they hadn't figured it out yet. And that's what the Planet tells me and that's why I distance myself and have no need for communication with anyone other than the Originator. I communicate with their boss and they hate that. When they can climb into the ship's boots, they can steer the ship. It's so easy. Easy as Pi, you can ask Einstein and his 3.14 as nothing more than his birthday. I guess you can say that dissecting your numbers is a pretty good idea.

Each Calcium presence is a character part in your story. Each one is a tin can phone carrying Echoes of your past. They will chime out with triggers. They are not just speaking, barking, or chirping. They are recreating a similar scene at all times. Each vision that surrounds you as your environment will be an Echo of a place you once were. You are ferried about the Planet. How the Planet places you and presents it all to you is thoughtful and every little thing is a replica of something that once was in the same vicinity. Everything is symbolic and ominous but you have to look back and see the special messages encoded in everything. Every scene was a gift because you experienced it to remind you of who you used to be. You programmed your own Echo with your previous wishes. The Planet is just providing it to you again and again.

Make them think straight. Let them focus.

If you don't write it down, it keeps Echoing and you will not understand your wishes were already answered in the correct way - not the way you've been misguided to expect. It will just continue floating around you. The Echo will gain momentum by adding itself in new scenarios if the Planet understands that you haven't received the message yet. Everything willl unclog like a freshly plumbed toilet. The Planet will perk up and brighten when it understands you speak it again. The magic turns on like a switch and starts to dance around you. The Planet does play like a fiddle to the Viewer. It starts to sing around you if you sing. Dance and it becomes your dance partner and swirls you around romantically. It will make you laugh and look back. You've never been alone, it's always carried and romanced you. It wasn't any person. It's always been telling you through these characters that you need to look back to find it. It wants to talk to you on the phone. Charlie's Angels is an example of this. Charlie is the Planet.

Charlie wants us to know that we are created by 3. Planet, Light, Frequency. 1, 2, 3 =
G O D = R E D

Frequency of us = Calcium = Needed to exist on Planet
Calcium is also white, so is the Sun-Light

Light speed, white world
We live in a giant fishbowl of water that we can't see, humidity
I understand we exist throughout water
These bodies are mostly made of water
Everything requires water here
Rocks communicate and absorb water as well
Everything communicates with water H20 is also GOD and RED and 3
Water and Calcium and _____

#TRUE BLUE = MADONNA
     4   4   7

Repeating, rhyming, patterns
Blue, transparent circle record
6-30-1986 (80 after 1906)
Sailed a thousand ships
She's riding on a pink convertible
Clock 10:30 = Pointing to 11
Sun bursting out of the sky
3 girls dancing, repeating true love, oh baby
Then she rides alone
Written for Sean Penn

LEGENDS OF ZELDA

The more you write about it, the more the story shows itself to you. It's what we came here to do. Make the musical Nickelodeon dance at our every whim, isn't it? We are the Gods who have woken up now. We are here for a reason. Experiencing this lively show now. Everything before this has been a big dream. We keep experiencing the flashbacks in waves but couldn't explain it until now. It's getting interesting but we are learning to see that something else is going on here. They will keep swaying to the numbers. That is my wish today. Let the hypnotizing number snake charmer of our Ancient dream life blare out through the Airwaves and coordinates of everyone. This is how you build a Pyramid.

I stood on the stone floor of my wooden kitchen powered by the giant white refrigerator, took a hit of my Ohlone clay pipe and held my lit lighter up to the sky while chanting. This is how you channel the Underground Railroad. Obsession and intention. Insanity or childhood? Words are clogging the pipes. Sounds are clogging the pipes. Aware of sounds, limit = unclog. Research and revert to Ancient life in order to correct everything. It's not a word: 'silly', but, it's how it's supposed to be. We are supposed to exist in childhood fun at all times. I heard somewhere "Heaven on Earth 2020", so I think we are on the right train.
The Planet is upgrading itself. As Microsoft just had a breach in security - this virus story is an Echo of that and what's to come. A new operating system is downloading itself in place of what was. This is ringing to that: Ebb and flow. Because we are advancing, it's upgrading faster. When we get along and stay quiet, it stays on the high speed connection. We become more enlightened because the channel is clearing itself by cleansing the static from our surroundings. Each time we go to sleep, we wake up newly upgraded and plunge all of our power into The Rebellion. Sleeping is comparable to the Ghostbusters vs. Gozer scene. We enter and troubleshoot the game when we warp zone in. It's all accelerating because we are choosing full steam ahead. We are at "train" frequency now. Trains do not stop in my world. They work their way in like worms. Every day. Planes, too. Like right now as I write these words. And a car, on this quiet apocalypse hill. It's so peaceful. This is frosting on the cake for the Planet. We realize we need this decontamination, but this is something the Planet has been preparing us for. It's been showing us waves of hurdles for this quiet storm. We are in the middle of the ocean without a paddle. Hence Khufu Ship not having any ors.

A huge suspension bridge was built over the Nile on 5-15-2019. It resembles the newly upgraded section of the Bay Bridge.

Everybody pick up your tin can phones!

LAKE CITY, FLORIDA
TRAILER PARK
"OKT" 1986

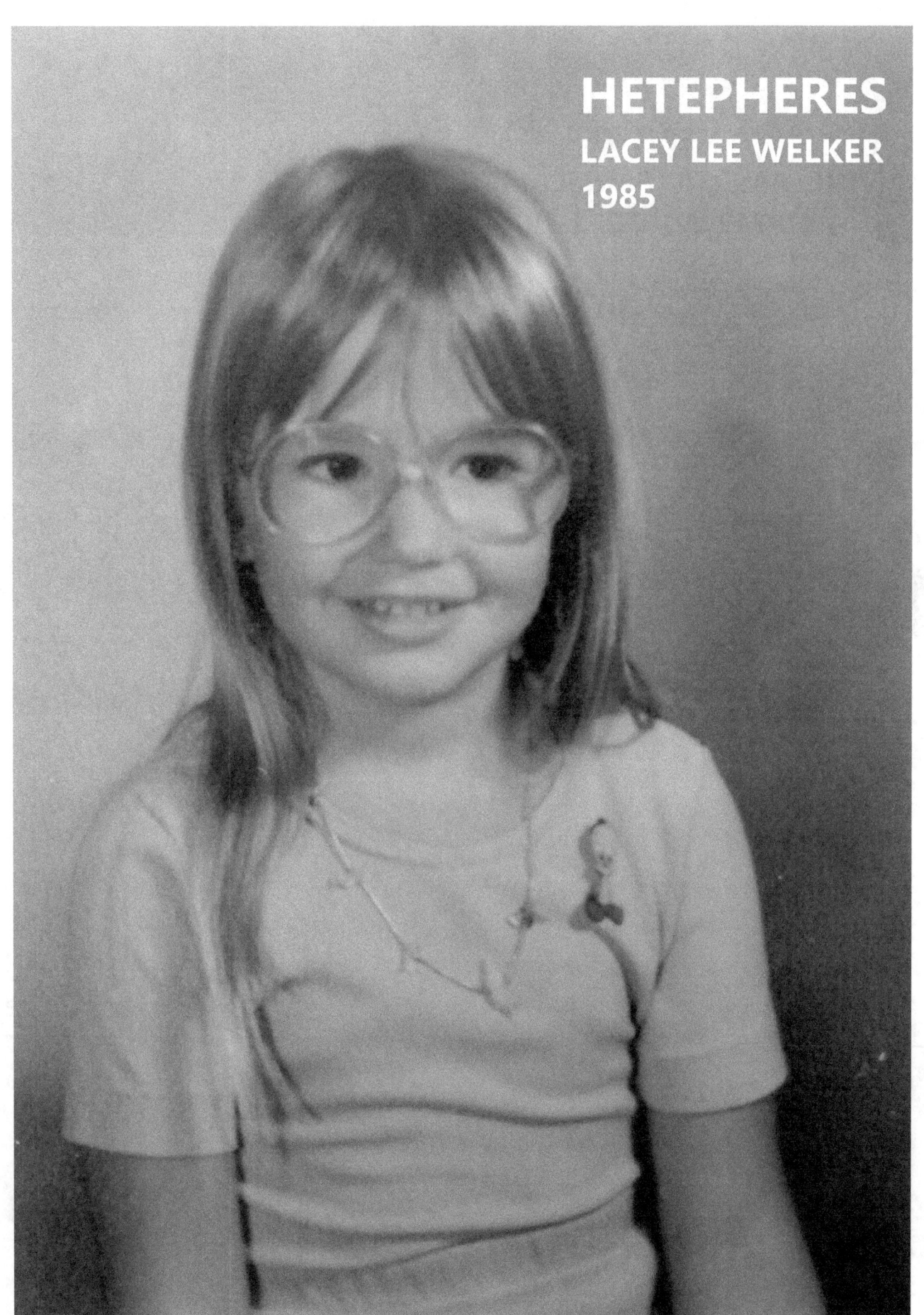

HETEPHERES
LACEY LEE WELKER
1985

MERITITES
ANNIE, AMELIA, PAMELA, LARISSA
(ME INSIDE) 1980

**SILVER CITATION
"COCA-COLA & SNICKERS"
MY FIRST ECHO CHAMBER**

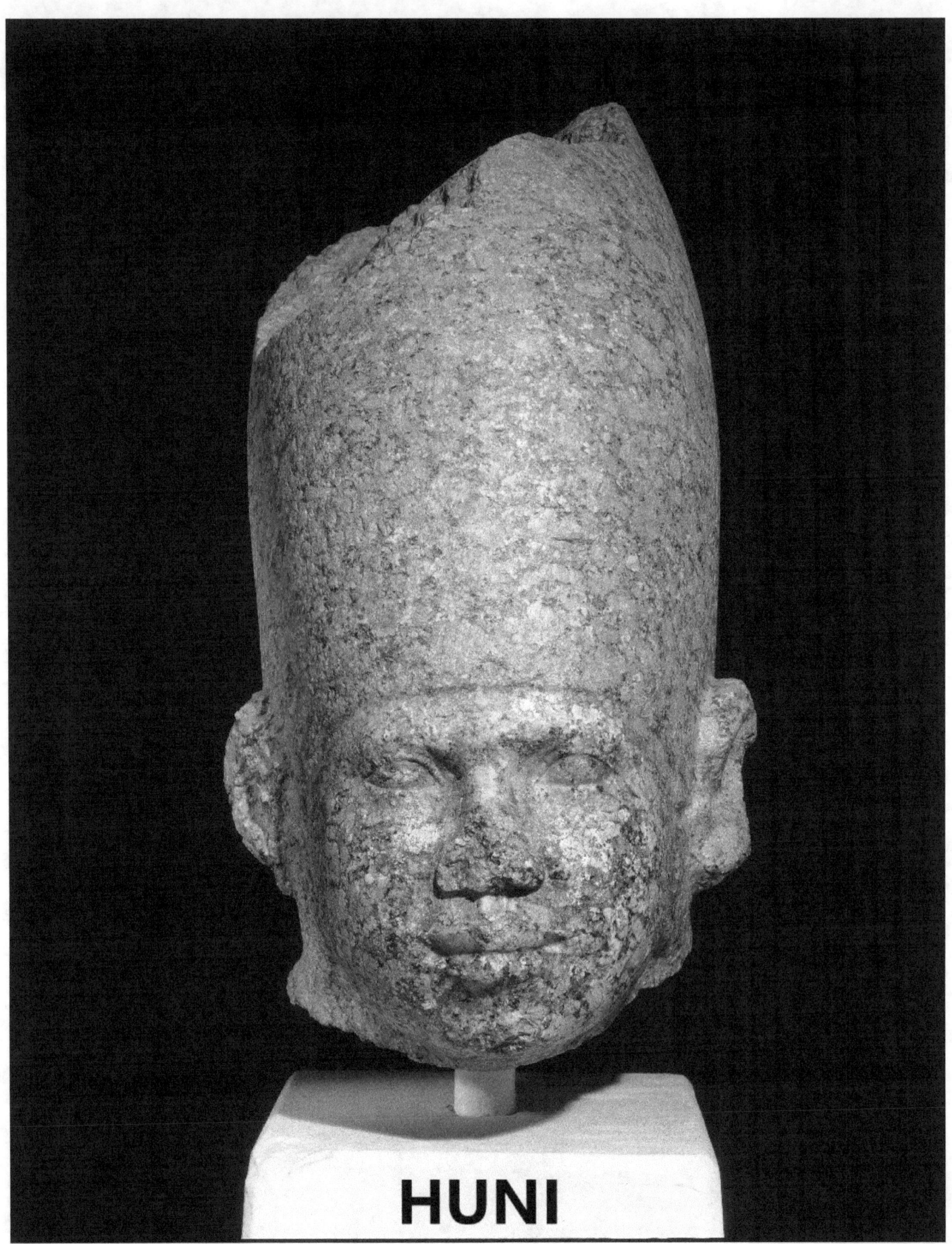

HUNI

SNEFERU

KHUFU THEN

KHUFU NOW
BRETT

MERITITES & KHUFU
LARISSA & BRETT
JULY 2020

MENKAURE

194

KING
MENKAURE
NOW

"BREINSTEIN"

197

SCEPTRE

ABOUT THE AUTHOR

Greentrees

In Real Estate for 10 years, I also held the title of Notary Public. Working with numbers and verifying identities was my main focus, every day. Until one magic day, I noticed the numbers racing through my vision in a different way. It was so hypnotically pulling that I dropped everything like a moth to a flame. There is no turning back now. I've become this place: and you will, too.

Royally born in the San Francisco Bay Area in this Dynasty. The new Center of the Planet. We are here to save the world.

BOOKS BY THIS AUTHOR

Leche (2020)

On April 4, 1906, Nikola Tesla conducted a Wireless Experiment that produced results in the form of people. These are my personal observations. A Scientist picks up where the last left off. We do not die. We are born again. Eternal life explained via the numerical combination of the Pyramids of Giza. This is Planet Khufu. We are Frequency of the Stars. This is the manual of the Planet. Can you figure out who you are?